Ceramic and Glass Materials

Structure, Properties and Processing

James F. Shackelford • Robert H. Doremus
Editors

Ceramic and Glass Materials

Structure, Properties and Processing

 Springer

Editors
James F. Shackelford
University of California, Davis
Dept. Chemical Engineering
 & Materials Science
1 Shields Avenue
Davis, CA 95616

Robert H. Doremus
Materials Research Center
Rensselaer Polytechnic Institute
Dept. Materials Science & Engineering
110 8th Street
Troy, NY 12180-3590

ISBN 978-0-387-73361-6   e-ISBN 978-0-387-73362-3
DOI 10.1007/978-0-387-73362-3

Library of Congress Control Number: 2007938894

Printed on acid-free paper

9 8 7 6 5 4 3 2 1

springer.com

Robert H. Doremus – A Dedication

With sadness, I note that in late January 2008 while finishing the editing of this book, Bob Doremus passed away suddenly in Florida. His wife and one of his daughters were with him at the time. Characteristic of his meticulous attention to detail, he had just finished personally preparing the index for this volume. Professor Doremus was an icon of ceramic and glass science, and this volume is a fitting tribute to his career. In addition to editing the book, he provided the opening chapter on alumina, the quintessential structural ceramic material.

After finishing *two* Ph.D. degrees in physical chemistry (University of Illinois, 1953 and University of Cambridge, 1956), Dr. Doremus worked at the General Electric Research and Development Laboratory for many years during a period of time that can fairly be described as a "golden age" of ceramic and glass science. His colleagues included Robert Coble, Joseph Burke, and Paul Jorgensen. There, he conducted seminal research including classic studies of gas and water diffusion in ceramics and glasses.

In 1971, he moved to the Department of Materials Science and Engineering at the Rensselaer Polytechnic Institute and began a long career as an educator. He continued to work on a broad range of topics in ceramic and glass science and was especially well known for publishing the definitive version of the important alumina-silica phase diagram [Klug, Prochazka, and Doremus, *J.Am.Ceram.Soc.*, **70** 750 (1987)] as well as doing pioneering work on bioceramics for medical applications. At Rensselaer, Bob was named the New York State Science and Technology Foundation Professor of Glass and Ceramics and served as Department Chair from 1986 to 1995.

Appropriate to his distinguished career as a scientist and educator, Bob received numerous awards in recognition of his accomplishments. Resulting in nearly 300 publications, his research contributions were recognized with the Scholes Award of Alfred University, the Morey Award of the American Ceramic Society, and the Ross Coffin Purdy Award, the American Ceramic Society's top honor for research. He received numerous teaching awards while at Rensselaer, including the Outstanding Educator Award of the American Ceramic Society. His winning the top research *and* educator awards of the American Ceramic Society is symbolic of his remarkable career.

Beyond these professional accomplishments of a great scientist and dedicated teacher, Bob Doremus was a devoted family man and leaves behind his wife Germaine and children Carol, Elaine, Mark, and Natalie. As with his family, Bob cared deeply about his students and worked tirelessly to help them. He was also a fine and supportive colleague. He will be greatly missed, and this book is dedicated to him with both affection and respect.

J.F. Shackelford
Davis, CA
February 2008

Preface

This book is intended to be a concise and comprehensive coverage of the key ceramic and glass materials used in modern technology. A group of international experts have contributed a wide ranging set of chapters that literally covers this field from A (Chap. 1) to Z (Chap. 10). Each chapter focuses on the structure–property relationships for these important materials and expands our understanding of their nature by simultaneously discussing the technology of their processing methods. In each case, the resulting understanding of the contemporary applications of the materials provides insights as to their future role in twenty-first century engineering and technology.

The book is intended for advanced undergraduates, graduate students, and working professionals. Although authored by members of the materials science and engineering community, the book can be useful for readers in a wide range of scientific and engineering fields.

Robert Doremus of the Rensselaer Polytechnic Institute covers one of the most ubiquitous modern ceramics in Chap. 1. The popularity of alumina by itself and as a component in numerous ceramic and glass products follows from its wide range of attractive properties. In Chap. 2, Duval, Risbud, and Shackelford of the University of California, Davis, look at the closely related and similarly ubiquitous material composed of three parts of alumina and two parts of silica, the only stable intermediate phase in the alumina–silica system at atmospheric pressure. Mullite has had significant applications in refractories and pottery for millennia and new applications in structures, electronics, and optics are the focus of active research. Richard Bradt of the University of Alabama, Tuscaloosa, provides Chap. 3, a focused discussion of the intriguing minerals (andalusite, kyanite, and sillimanite) that do not appear on the common alumina–silica phase diagram as they are formed at high geological pressures and temperatures. Nonetheless, these minerals with a one-to-one ratio of alumina to silica are widely found in nature and are used in numerous applications such as refractories for the steel and glass industries. In Chap. 4, Martin Wilding of the University of Wales, Aberystwyth, further expands the compositional range of materials considered by exploring the ceramics and glasses formed in binary aluminate systems. Sharing the high melting point and chemical resistance of the alumina end-member, these aluminates find a wide range of applications from cements to bioceramics and electronic components.

In Chap. 5, Davila, Risbud, and Shackelford of the University of California, Davis, review the various ceramic and glass materials that come from silica, the most abundant mineral in the Earth's crust. The many examples they give share a simple chemistry but display a wide range of crystalline and noncrystalline structures. The materials also represent some of the most traditional ceramic and glass applications as well as some of the most sophisticated, recent technological advances. In Chap. 6, Smith and Fahrenholtz of the University of Missouri, Rolla, cover a vast array of ceramic materials, including many of the materials covered in other chapters in this book. The resulting perspective is useful for appreciating the context in which ceramics are used for one of their most important properties, viz. the resistance to high temperatures. Professor Fahrenholtz then provides a comprehensive coverage of clays in Chap. 7. These important minerals that serve as raw materials for so many of the traditional ceramics are also providing a framework for the science of the study of advanced ceramics. In Chap. 8, Mariano Velez of the Mo-Sci Corporation reviews the ceramic oxides that are used for the two distinctive markets of (a) structural applications and (b) high temperature (refractory) concretes.

Professor Julie Schoenung of the University of California, Davis, reviews a wide range of minerals in Chap. 9. These materials produce the various lead oxides and silicates so widely used in lead-containing glasses and crystalline electronic ceramics. The regulatory issues surrounding these well known carcinogenic materials are also discussed. Finally in Chap. 10, Olivia Graeve of the University of Nevada, Reno, reviews the complex structural and processing issues associated with the family of ceramics zirconia that is widely used because of the superior values of toughness and ionic conductivity.

Finally, we thank the staff of Springer for their consistent encouragement and professional guidance in regards to this book. We especially appreciate Gregory Franklin for helping to initiate the project and Jennifer Mirski for guiding it to completion.

Davis, CA
Troy, NY

Shackelford
Doremus

Contents

Contributors

Richard C. Bradt
Department of Materials Engineering, The University of Alabama, Tuscaloosa, AL 35487-0202, USA, rcbradt@eng.ua.edu

Lilian P. Davila
Department of Chemical Engineering and Materials Science, University of California, Davis, CA 95616, USA, lpdavila@ucdavis.edu

Robert H. Doremus
Department of Materials Science and Engineering, Rensselaer Polytechnic Institute, Troy, NY, USA

David J. Duval
Department of Chemical Engineering and Materials Science, University of California, Davis, CA 95616, USA, djduval@ucdavis.edu

William G. Fahrenholtz
Materials Science and Engineering Department, Missouri University of Science and Technology, Rolla, MO 65409, USA, billf@mst.edu

Olivia A. Graeve
Department of Chemical and Metallurgical Engineering, University of Nevada, Reno, NV, USA, oagraeve@unr.edu

Subhash H. Risbud
Department of Chemical Engineering and Materials Science, University of California, Davis, CA 95616, USA, shrisbud@ucdavis.edu

Julie M. Schoenung
Department of Chemical Engineering and Materials Science, University of California, Davis, CA 95616, USA, jmschoenung@ucdavis.edu

James F. Shackelford
Department of Chemical Engineering and Materials Science,
University of California, Davis, CA 95616, USA, jfshackelford@ucdavis.edu

Jeffrey D. Smith
Materials Science and Engineering Department, Missouri University of Science and
Technology, Rolla, MO 65409, USA, jsmith@mst.edu

Mariano Velez
Mo-Sci Corporation, Rolla, MO 65401, USA, mvelez@mo-sci.com

Martin C. Wilding
Institute of Mathematical and Physical Sciences, University of Wales,
Aberystwyth, Ceredigion SY23 3BZ, UK, mbw@aber.ac.uk

Chapter 1
Alumina

Robert H. Doremus

The uses, processing, structure, and properties of alumina are summarized in this article. Various polymorphs of alumina and its phase relations with other oxides are described. The following properties are discussed: mechanical, thermal, thermodynamic, electrical, diffusional, chemical, and optical. Quantitative values for these properties are given in tables. The usefulness of alumina results from its high strength, melting temperature, abrasion resistance, optical transparency, and electrical resistivity. Traditional uses of alumina because of these properties are furnace components, cutting tools, bearings, and gem stones; more recent applications include catalyst substrates, tubes for arc lamps, and laser hosts. Possible new uses of alumina are in electronic circuits, optical components, and biomaterials. Alumina fibers for composites and optics must be pure, defect free, and cheap.

1 Introduction

Alumina (Al_2O_3) is one of the most important ceramic materials, both pure and as a ceramic and glass component. Some uses of alumina are given in Table 1; an exhaustive and detailed description of many of these uses is given in [1]. There are also extensive discussions of uses of alumina in [2]. Processing of alumina is discussed in both these references, and [2] has a summary of some properties of alumina. In writing this review, I have relied on material from these references. Anyone interested in more details of processing and properties of alumina should obtain [2] from the American Ceramic Society. Reference [1] is also available from the Society and has additional information on processing and uses of alumina. More recently, there have been two issues of the Journal of the American Ceramic Society devoted to alumina [3, 4]; these issues concentrate on defects and interfaces, especially grain boundaries [3], grain growth, and diffusion in alumina [4].

The usefulness of alumina derives from a variety of its properties. It has a high melting temperature of 2,054°C, and is chemically very stable and unreactive, leading to applications as high-temperature components, catalyst substrates, and biomedical implants. The hardness, strength, and abrasion resistance of alumina are among the highest for oxides, making it useful for abrasive materials, bearings,

J.F. Shackelford and R.H. Doremus (eds.), *Ceramic and Glass Materials:*
Structure, Properties and Processing.
© Springer 2008

Table 1 Uses of Alumina

Solid alumina

Furnace components
Catalyst substrates
Electronics substrates
Electrical insulators
Cutting tools
Bearings
Spark Plugs
Arc lamp tubes
Laser hosts
Gem stones

Alumina powders

Abrasives
Catalyst pellets

Alumina coatings

Oxidation protection of aluminum
    and aluminum alloys
Capacitors
Transisitors
Bioceramics

Alumina fibers

Thermal insulators
Fire retardation

Alumina as a component of

Ceramics and glasses
Mullite components
Electrical insulators
Porcelains
Durable glasses

and cutting tools. The electrical resistance of alumina is high, so it is used pure and as a component in electrical insulators and components. Alumina has excellent optical transparency, and along with additives such as chromium and titanium, it is important as a gem stone (sapphires and rubies) and a laser host (ruby). Because of its high melting temperature, chemical inertness, and optical transparency, it is highly useful for containing arcs in street lamps. See Table 1 and also [1, 2] for more on uses of alumina.

In this review, the processing of alumina is discussed next, and then its properties are tabulated and described. In a summary future uses of alumina are considered.

2 Processing

2.1 *Raw Materials*

Bauxite is the name of the ore that is the primary source of alumina; bauxite contains gibbsite, γ-Al(OH)$_3$, which is the stable phase of Al(OH)$_3$ at ambient temperature and pressure. The structure of alumina and hydrated alumina phases are listed in Table 2.

Table 2 Structures of stable alumina (corundum) and unstable aluminas

Designation	Structure	Lattice Parameters, angle (Å)		
		a	b	c
Corundum	Hexagonal (rhombohedral)	4.758		12.991
Eta	Cubic (spinel)	7.90		
Gamma	Tetragonal	7.95		7.79
Delta	Tetragonal	7.97		23.47
Theta	Monoclinic	5.63	2.95	11.86 103° 42′
Kappa	Orthorhombic	8.49	12.73	13.39

Bauxite from the Guianas in South America is low in iron and silica impurities, so it is preferred when purity is important. Other important sources of bauxite are in Brazil, the southern United States, Southeast Asia, West Africa, and India.

2.2 Processing

Aluminum hydroxides are separated from bauxite by the Bayer process, in which these hydroxides are dissolved in sodium hydroxide to separate them from the other unwanted constituents of the bauxite. The dissolution reactions are carried out at about 285°C and 200 atm. pressure, and are:

$$Al(OH)_3(s) + NaOH(soln) = NaAl(OH)_4(soln) \qquad (1)$$

$$AlOOH(s) + H_2O(soln) + NaOH(soln) = NaAl(OH)_4(soln) \qquad (2)$$

in which (s) stands for solid and (soln) for solution. The solution containing $NaAl(OH)_4$ is separated from the unwanted solid impurities by sedimentation and filtration, and the solute is cooled to about 55°C. The aluminum hydroxides precipitate from the solution, aided by the addition of gibbsite seeds. The dried precipitated alumina or aluminum hydroxides can be used directly or further purified by resolution and reprecipitation. Other methods for preparing alumina and aluminum hydroxides from bauxite are described in [1, 2].

The stable phase of alumina at all temperatures and ambient pressure (one atm, or $(1.01)\ 10^5$ Pa) is corundum or α-Al$_2$O$_3$ (see Table 2). In single-crystal form, corundum is called sapphire. No phase transformation of corundum up to 175 GPa pressure has been observed experimentally [5, 6]; however, a calculation predicts that corundum should transform to the Rh_2O_3 (II) structure at about 78 GPa, and to a cubic perovskite structure at 223 GPa [7]. The Rh_2O_3 (II) structure has an X-ray pattern close to the corundum structure, so the transformation may have been missed in experimental studies.

Solid polycrystalline alumina is made from alumina powder by sintering. The traditional sintering methods for ceramics involve forming a powder into "green" ware, partially drying it at low temperatures, possibly "calcining" (heating) it at intermediate temperatures (perhaps 900°–1,100°C), and firing it to a dense solid at high temperature, for alumina above 1,400°C.

The time and temperature required to form the desired degree of porosity in the dense solid depend mainly on the particle size of the alumina powder. The usual sintering sequence is imagined to be: (1) neck formation between powder particles, (2) formation of open porosity with a continuous solid phase (intermediate stage), and (3) removal of closed pores imbedded in the dense solid. In the usual practical sintering of alumina, stage one is rapid, and the final density or porosity is determined mainly by stage three.

Various other oxides have been added to alumina to reduce the porosity of the final sintered solid. An especially valuable finding by Coble [9] was to add MgO to pure alumina powder; the resulting sintered alumina can be translucent (partial transmission of light). Usually sintered ceramics are opaque because of light scattering from residual pores, but in the translucent alumina, called Lucalox™, the porosity is low enough to reduce this scattering, so that Lucalox tubes are used in street lamps for containing a sodium arc. Because the lamp can be operated at high temperature, it is quite efficient.

Dense alumina can also be made by melting, but the high-melting temperature of 2,054°C makes this process expensive and difficult to control. High-value materials such as gem stones and laser hosts can be made by adding various colorants such as chromium, titanium, iron, cobalt, and vanadium to the melt.

In the sintering of alumina powders, the desired shape is formed in the green state before drying and firing. Various other constituents can be added to the starting powder. The density of the final product can be increased by hot-pressing, that is, by carrying out the firing under pressure. This method is expensive, so it is used only for high value polycrystalline products.

Alumina refractories for use in high temperature applications such as glass melting furnaces are usually made by the fusion-cast process. Various other oxides, such as SiO_2, MgO, Cr_2O_3, and ZrO_2 are added to the alumina powder to lower its melting point, and the resulting mixture is melted in an electric furnace and cast into the desired shapes for refractory applications. See the section on phase diagrams for the melting temperatures and compositions of a few mixtures of alumina with other oxides.

3 Structures of Pure and Hydrated Alumina

The structures of aluminas and hydrated aluminas are given in Tables 2 and 3. The only stable phase of Al_2O_3 is corundum at all temperatures and up to at least 78 GPa pressure (see earlier discussion). The corundum structure is shown in Fig. 1. It consists of oxygen ions in a slightly distorted close-packed hexagonal (rhombohedral) lattice, space group R3c. The aluminum ions occupy two-thirds of the octahedral sites in the oxygen lattice. The lattice parameters for corundum in Table 2 are for a hexagonal unit cell containing 12 Al_2O_3 molecules. The rhombohedral lattice parameters are $a = 5.128$ Å and $\alpha = 55.28°$. The ionic porosity Z of a solid is given by the formula

$$Z = 1 - V_a/V \qquad (3)$$

in which V_a is the volume of atoms in a molecule (or in the unit cell) and V is the specific volume, or the volume of the unit cell. For alumina $Z = 0.21$ with the radius

Table 3 Structures of hydrated alumina phases

Phase	Formula	Lattice parameters (Å)/angle		
		A	B	c
Bayerite β-Al(OH)$_3$	Monoclinic	4.72	8.68	5.06/90°7′
Gibbsite α-Al(OH)$_3$	Monoclinic	8.64	5.07	9.72/85°26′
Boehmite α-AlOOH	Orthorhombic	2.87	12.23	3.70
Diaspore β-AlOOH	Orthorhombic	4.40	9.43	2.84

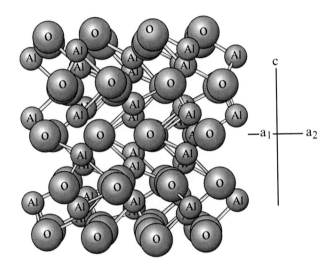

CORUNDUM: Al$_2$O$_3$

Fig. 1 The structure of alpha alumina from [31]. The structure of corundum (alpha-alumina) from [31]. The aluminum atoms occupy two-thirds of the octahedral interstices in a hexagonal close-packed array of oxygen atoms, which is distorted because the octahedral share faces in pairs

of 1.38 Å for oxygen atoms, so the structure has less "open" volume even than close-packing ($Z = 0.26$) of uniform spheres.

The various metastable alumina structures are all less dense than corundum. Several other allotropic structures have been suggested, but are less well-verified than those in Table 2. All these metastable aluminas have oxygen packings that are near to close-packed cubic. Usually, eta or gamma aluminas are formed at low temperatures, and transform in the sequence gamma→delta→theta→alpha alumina with increasing temperatures. However, many other variants are possible, with gamma formed at higher temperatures and transforming directly to alpha. See [8] for some previous references. Factors such as particle size, heating rate, impurities, and atmosphere can influence the kinetics of transformation and the sequence of phases. Above about 1,200°C, only alpha phase (corundum) is usually present.

The structures of various hydrated aluminas are given in Table 3. One configuration suggested for these structures is chains of Al–O bonds with hydrogen bonding between chains. These hydrated aluminas decompose at low temperature (about 300°C) to Al$_2$O$_3$ and water.

4 Equilibrium Binary Phase Diagrams of Alumina with Other Oxides

The most important binary oxide and ceramic phase diagram is the alumina–silica (Al_2O_3–SiO_2) diagram, shown in Fig. 2, as determined by Klug [10]. Important features in this diagram are the very low solid solubility of SiO_2 in Al_2O_3 and Al_2O_3 in SiO_2 and the single stable intermediate solid phase of mullite, which has the composition $3Al_2O_3$–$2SiO_2$; at higher temperatures, the amount of alumina in mullite increases. In contrast to binary metal systems, which usually have considerable solid solubility in the pure components and limited solubility in intermetallic phases, there is some solid solubility in mullite and very little in the end members of SiO_2 (cristobalite) and Al_2O_3 (corundum).

There is complete solid solubility in the system Al_2O_3–Cr_2O_3; both end members have the corundum structure [11, 12]. There is also subsolidus phase separation in this system [13]. There is also considerable solid solubility in the end-member oxides in the Al_2O_3–Fe_2O_3, Al_2O_3–Y_2O_3, and Al_2O_3–Ga_2O_3 systems [12]. Thus, the solid solubility results because the three-valent ions of Cr, Fe, Y, and Ga can substitute for aluminum in the corundum structure, and aluminum substitutes for these ions in their oxides. Alternatively, the solubilities of oxides with cation valences other than plus three are usually very low. For example, the solubility of magnesia in alumina is about 1 ppm atom fraction (Mg/Al) at 1,200°C [14]. Thus in mixtures of Al_2O_3 with higher concentrations of MgO than 1 ppm, second phases containing MgO can form on grain boundaries at 1,200°C and lower temperatures, but the Mg atoms do not dissolve or substitute for Al in the Al_2O_3. In the literature, there are many reports of higher solubilities of ions with valences different from three in

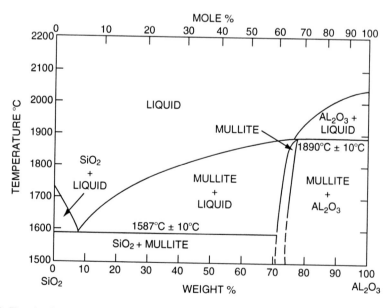

Fig. 2 The alumina–silica phase diagram. From [10]

alumina, but in view of the very careful work of Greskovitch and Brewer [14] with extremely pure alumina, these higher solubilities are unlikely.

In [1] there is a table (XI, on page 63) of minimum melting temperatures for a variety of binary alumina-oxide systems. With most oxides, these eutectic or peritectic temperatures vary from about 1,500 to 2,000°C. The Al_2O_3–V_2O_5 system has an anomalously low eutectic of 660°C at 99% V_2O_5; other low melting mixtures are Al_2O_3–Bi_2O_3 of 1,070°C, Al_2O_3–WO_3 of 1,230°C, and Al_2O_3–B_2O_3 of 1,440°C.

5 Mechanical Properties

5.1 Elasticity

Alumina shows the deformation behavior of a typical brittle solid, which is linear elasticity to failure. Elastic deformation is instantaneous when stress is applied, and is completely reversible when it is removed. In a tensile test of a rod or bar of alumina, the strain is linear with stress to failure; the slope of this stress–strain curve is the Young's modulus. Values of various moduli and Poisson's ratio for pure, dense polycrystalline alumina are given in Table 4 (from [15]). These values are considerably higher than those for most other oxides, as a result of the strong (high energy) aluminum–oxygen bonds in alumina. The various elastic constants of single crystal alumina are given in Table 5. As the temperature increases the elastic moduli decrease (as shown in Table 6) because of the increase in atomic displacements as the temperature increases, and consequent reduced bond strength.

Table 4 Elastic properties of polycrystalline alumina at room temperature [15], moduli in GPa

Young's modulus	403
Shear modulus	163
Bulk modulus	254
Poisson's ratio	0.23

Table 5 Elastic properties of single crystal alumina at room temperature [2, 15] in GPa

C11	498	S11	23.5
C12	163	S12	7.2
C13	117	S13	3.6
C14	−23.5	S14	4.9
C33	502	S33	21.7
C44	147	S44	69.4

Table 6 Temperature dependence of Young's modulus for polycrystalline alumina [2]

Temp. (°C)	Young's modulus (GPa)
25	403
500	389
1,000	373
1,200	364

5.2 Strength

The mechanical strength of a brittle material such as alumina depends on flaws (cracks) in the alumina surface. When a tensile stress is applied perpendicular to a deep, thin crack, the stress at the tip of the crack is greatly magnified above the applied stress. Thus, the surface condition of a brittle solid determines it strength. Surface flaws develop from abrasion, so the higher the abrasion resistance of a brittle solid the greater its practical strength. Strengths of alumina are given in Table 7.

If a solid has no surface or internal flaws (a "perfect" lattice), it should have very high strength. Various theoretical equations for this ultimate strength S of a brittle solid have been proposed; one is [16]

$$S^2 = E\gamma/4b \qquad (4)$$

in which E is Young's modulus, γ is the surface energy, and b the lattice parameter. With $E = 403\,GPa$ (Table 4), $\gamma = 6.0\ J\ m^{-2}$ [17, 18], and $b = 0.177\,nm$, the ultimate strength S of alumina is about 58 GPa. This value is very high because of the high bond strength of alumina; for example, silicate glasses and quartz have theoretical strength values of 18 GPa or lower.

Practical strengths of brittle materials vary over wide ranges depending on their surface condition and history. For alumina, tensile or bonding strengths vary over a wide range of values because of different surface conditions, resulting in different flaw depths and flaw distributions. See [19] for a discussion of flaw distribution functions. The strength values for alumina are higher than for most other oxides; of course all of these strengths are far smaller than the theoretical strength, and depend strongly on the history and treatment of the samples. As the temperature increases, the strength of alumina decreases (as shown in Table 7) because of the increase of atomic vibrations and reduction in bond strength, just as for the reduction in elastic modulus with temperature. The strength of polycrystalline alumina depends strongly on its grain size, as shown by one set of strength values from [2]. See also [20] for strengths of alumina machined and annealed at different temperatures. The strength also decreases as the alumina becomes more porous, as shown in Table 8; isolated pores increase the applied stress on their surfaces, and open porosity means much more surface for flaw development.

Table 7 Bend strengths of alumina in MPa

Theoretical strength 58,000 at 25°C				Ref.
Single crystals (sapphire) 300–700 at 25°C				2
Polycrystals with similar treatment,				2
as a function of grain size in micrometers:				
Grain size →	1–2	10–15	40–50	2
Temp (°C)				
25°C	460	330	240	
400°C	360	260	230	
1,000°C	340	260	210	
1,350°C	260	110	97	

Table 8 Effect of porosity on the bend strength of polycrystalline alumina at 25°C from [2]

Porosity (%)	Strength (MPa)
0	269
10	172
20	110
30	76
40	55
50	47

5.3 Fatigue

The strengths of crystalline and glassy oxides decrease with time under a constant applied load. This static fatigue is usually modeled with a power law equation between times to failure t when a sample is subjected to an applied stress s:

$$\log t = c - n\log s \tag{5}$$

in which c is a constant and the stress exponent n is a measure of the susceptibility of the material to fatigue. The larger the n value the more resistant the material is to fatigue. Typical values of n for silicate glasses are 13 or lower [21]; for alumina an n value of about 35 was found [21], showing that alumina has much better fatigue resistance than most other oxides under ambient conditions.

This fatigue in oxides results from reaction with water, which can break the cation–oxygen bonds in the material; for example in alumina:

$$Al - O - Al + H_2O = AlOH + HOAl \tag{6}$$

Thus, when the ambient atmosphere is dry the fatigue failure time is long, and as the humidity increases the fatigue time decreases.

5.4 Hardness

The hardness of a material is measured by pressing a rod tip into a material and finding the amount of deformation from the dimensions of the resulting indentation. Hardness measurements are easy to make but hard to interpret. The stress distribution under the indenter is complex, and cracking, elastic and anelastic deformation, faulting, and plastic deformation are all possible around the indentation. Alumina is one of the hardest oxides. On the nonlinear Mohs scale of one to ten, alumina is nine and diamond is ten, but diamond is about a factor of three harder than alumina. Some approximate Knoop hardness (elongated pyramidal diamond indenter) values for alumina are given as a function of temperature in Table 9, and in Table 10 for some hard ceramics [22, 23]. It is curious that the hardness of alumina decreases much more than the strength as the temperature is increased.

Table 9 Knoop hardness of alumina as a function of temperature

T (K)	Hardness (kg mm^{-2})
400	1,950
600	1,510
800	1,120
1,000	680
1,200	430
1,400	260
1,600	160

Table 10 Knoop hardness values of some ceramics at 25°C from [2, 22]

Material	Hardness (kg mm^{-2})
Diamond	8,500
Alumina	3,000
Boron carbide	2,760
Silicon carbide	2,480
Topaz ($Al_{12}Si_6F_{10}O_{25}$)	1,340
Quartz (SiO_2)	820

5.5 Creep

Creep is the high-temperature deformation of a material as a function of time. Other high-temperature properties related to creep are stress and modulus relaxation, internal friction, and grain boundary relaxation. The creep rate increases strongly with temperature, and is often proportional to the applied stress. Microstructure (grain size and porosity) influences the creep rate; other influences are lattice defects, stoichiometry, and environment. Thus, creep rates are strongly dependent on sample history and the specific experimental method used to measure them, so the only meaningful quantitative comparison of creep rates can be made for samples with the same histories and measurement method. Some torsional creep rates of different oxides are given in Table 11 to show the wide variability of creep values. Compared with some other high temperature materials such as mullite ($3Al_2O_3 \cdot 2SiO_2$), alumina has a higher creep rate, which sometimes limits its application at high temperatures (above about 1,500°C). See [24] for a review of creep in ceramics and [25] for a review of creep in ceramic–matrix composites.

5.6 Plastic Deformation

At high temperatures (above about 1,200°C) alumina can deform by dislocation motion. The important paper by Merritt Kronberg [26], see also [1], p. 32, and [27], showed the details of dislocation motion in alumina. Basal slip on the close-packed oxygen planes is most common in alumina, with additional slip systems on prism planes.

Table 11 Torsional creep rates of some polycrystalline oxides at 1,300°C and 124 MPa applied stress (from [23], p. 755)

Material	Creep rate ($\times 10^5\,h^{-1}$)
Al_2O_3	0.13
BeO	30
MgO (slip cast)	33
MgO (pressed)	3.3
$MgAl_2O_4$, spinel (2–5 µm grains)	26
$MgAl_2O_4$ (1–3 µm grains)	0.1
ThO_2	100
ZrO_2	3

5.7 Fracture Toughness

The fracture toughness K_{IC} of a brittle material is defined as

$$K_{IC} = YS\sqrt{c} \tag{7}$$

in which S is the applied stress required to propagate a crack of depth c, and Y is a geometrical parameter. Values of K_{IC} are often measured for ceramics from the lengths of cracks around a hardness indent. K_{IC} is not material parameter; it depends on sample history and many uncontrolled factors. It is based on the Griffith equation, which gives a necessary but not sufficient criterion for crack propagation [18]. Thus K_{IC} is not a very useful quantity for defining mechanical properties of brittle material. A value of about 3.0 MPam$^{1/2}$ is often found for alumina [1].

6 Thermal and Thermodynamic Properties

6.1 Density and Thermal Expansion

The density of alpha alumina at 25°C is 3.96 g cm^{-3}, which gives a specific volume of 25.8 cm^3 mol^{-1} or 0.0438 nm^3 per Al_2O_3 molecule. Densities of other aluminas are given in Table 12.

The coefficient of thermal expansion α of alumina at different temperatures is given in Table 13. Often an average value of α is given over a range of temperatures, but the slope of a length vs. temperature plot at different temperatures is a more accurate way of describing α.

6.2 Heat Capacity (Specific Heat) and Thermodynamic Quantities

The specific heat, entropy, heat and Gibbs free energies of formation of alumina are given in Table 14, from [28]. Above 2,790°K, the boiling point of aluminum, there is a discontinuous change in the heat of formation of alumina.

Table 12 Densities of anhydrous and hydrated aluminas

Material	Density (g cm^{-2})
Sapphire (α-A$_{12}$O$_3$)	3.96
γ-Al$_2$O$_3$	3.2
δ-Al$_2$O$_3$	3.2
κ-Al$_2$O$_3$	3.3
θ-A$_{12}$O$_3$	3.56
Al(OH)$_3$ gibbsite	2.42
AlOOH diaspore	3.44
AlOOH boehmite	3.01

From [1]

Table 13 The coefficient of linear thermal expansion of α–alumina as a function of temperature

Temp. (°C)	$d\alpha/dT$ ($\times 10^6$ per °C)
1,000	12.0
800	11.6
600	11.1
400	10.4
200	9.1
100	7.7
50	6.5

From [23]

Table 14 Specific heat and thermodynamic properties of α-alumina as a function of temperature

T (K)	Specific heat	Entropy	Heat of formation	Free energy of formation
	J mol^{-1} K^{-1}		kJ mol^{-1}	
0	0	0	−1663.608	−1663.608
100	12.855	4.295	−1668.606	−1641.692
200	51.120	24.880	−1673.388	−1612.636
298.15	79.015	50.950	−1675.692	−1582.275
400	96.086	76.779	−1676.342	−1550.226
600	112.545	119.345	−1675.300	−1487.319
800	120.135	152.873	−1673.498	−1424.931
1,000	124.771	180.210	−1693.394	−1361.437
1,200	128.252	203.277	−1691.366	−1295.228
1,400	131.081	223.267	−1686.128	−1229.393
1,600	133.361	240.925	−1686.128	−1163.934
1,800	135.143	256.740	−1683.082	−1098.841
2,000	136.608	271.056	−1679.858	−1034.096
2,200	138.030	284.143	−1676.485	−969.681
2,327	138.934	291.914	Melting temperature	
2,400	139.453	296.214	−1672.963	−905.582
2,600	140.959	307.435	−1669.279	−841.781
2,800	142.591	317.980	−2253.212	−776.335
3,000	144.474	327.841	−2244.729	−671.139

From [28]

6.3 Vaporization of Alumina

There is a detailed discussion of the vaporization behavior of alumina and other oxides in [29]. The main vapor species over alumina are Al, AlO, Al_2O, and AlO_2, depending on the temperature and oxidizing or reducing conditions in the surrounding atmosphere. Under reducing conditions Al and Al_2O are predominant; in 0.2 bar O_2, both AlO and AlO_2 are the main species [30].

Two examples of quantitative data of vapor pressure as a function of temperature are given in Table 15.

The boiling temperature of alumina at one atm pressure is about 3,530°C with a heat of vaporization of about 1,900 kJ mol^{-1} at 25°C [2], when compared with the melting temperature of 2,054°C, and a heat of fusion of about 109 kJ mol^{-1} at 25°C [31].

6.4 Thermal Conductivity

The thermal conductivity of α-alumina single crystals as a function of temperature is given in Table 16 (from [2, 23]). Heat is conducted through a nonmetallic solid by lattice vibrations or phonons. The mean free path of the phonons determines the thermal conductivity and depends on the temperature, phonon–phonon interactions, and scattering from lattice defects in the solid. At temperatures below the low temperature maximum (below about 40°K), the mean free path is mainly determined by the sample size because of phonon scattering from the sample surfaces. Above the maximum, the

Table 15 The pressure of AlO vapor and total vapor pressure in equilibrium with α-Al_2O_3 as a function of temperature, for reducing and neutral conditions

Temp. (K)	Log vapor pressure of AlO, P (bar) [29]
1,520	−15
1,630	−13
1,750	−11
1,900	−9
2,020	−7
2,290	−5

Temp. (K)	Log total vapor pressure, P (atm.) [2, 31]
2,309	−5.06
2,325	−4.99
2,370	−4.78
2,393	−4.77
2,399	−4.66
2,459	−4.42
2,478	−4.24
2,487	−4.04
2,545	−3.70
2,565	−3.89
2,605	−3.72

Table 16 Thermal conductivity of single crystal α-Al_2O_3

Temp. (K)	Conductivity (J s^{-1}mK^{-1})	Temp. (°C)	Conductivity (J s^{-1}mK^{-1})
0	0	25	36
10	1,200	100	29
20	3,800	300	16
40	5,900	500	10
50	5,000	700	7.5
60	2,300	900	6.3
80	790	1,100	5.9
100	400	1,300	5.9
200	100	1,500	5.4
		1,700	5.9
		1,900	6.3

From [2, 23]

conductivity decays approximately exponentially because of phonon–phonon interactions. At high temperatures (above about 800°C), the phonon mean free path is of the order of a lattice distance, and becomes constant with temperature. There is a much more detailed discussion of phonon behavior in ceramics and glasses in [23, 32]. The velocity v of a phonon or sound wave in a solid can be found from the formula

$$v^2 = E/\rho \tag{8}$$

in which E is Young's modulus and ρ is the density, so this velocity in alumina is $10.1(10)^3$ m s^{-1} at 25°C. This result is close to the measured value of 10.845 m s^{-1}.

7 Electrical Properties

7.1 Electrical Conductivity

There have been a large number of studies of electrical conductivity of alumina, with widely different values being reported. Papers before 1961 are listed in [33] and those from 1961 to 1992 in [34].

Anyone interested in the electrical conductivity of alumina should read carefully the papers of Will et al. [34]. These authors measured the electrical conductivity of highly pure and dry sapphire from 400°C to 1,300°C; the elemental analysis of their sapphire samples is given in Table 17, and showed less than 35 ppm total impurities. Particularly significant is the low level of alkali metal impurities, which often provide ionic conduction in oxides.

The measurements in [34] were made with niobium foil electrodes with a guard ring configuration on disc samples, and in a vacuum of 10^{-7}–10^{-8} Torr. A nonsteady-state voltage sweep technique was used for the measurements. The results are in Table 18 and Fig. 3 for conductivity along the x-axis. Between 700°C and 1,300°C, the activation energy was about 460 kJ mol^{-1} (4.8 eV) and between 400 and 700°C it was 39 kJ mol^{-1} (0.4 eV). The great care taken with these measurements and the high purity of the sapphire make them definitive for the electrical conductivity for pure, dry alumina.

Table 17 Chemical analysis of sapphire for electrical conductivity measurements, from [34]

Element	Conc. (ppm)	Element	Conc. (ppm)
Iron	8	Potassium	<5
Silicon	6	Sodium	<3
Calcium	3	Nickel	<3
Magnesium	0.6	Chromium	<3
Beryllium	0.1	Lithium	<2

Table 18 Electrical conductivity of pure, dry sapphire

Temp. (°C)	Log conductivity $(ohm^{-1} cm^{-1})$
1,300	7.46
1,200	8.48
1,100	9.70
1,000	11.14
900	12.88
800	14.24
700	15.20
600	15.32
500	15.70
400	16.08

From [34]

After 650 h electrolysis at 1,200°C, the conductivity remained constant, showing it was electronic and nonionic [34]. The authors [34] interpreted their results in terms of electrical conductivity of a wide-band semiconductor. The high-temperature portion resulted from intrinsic conductivity with equal numbers of holes and electrons as carriers; twice the activation energy gives the band gap of about 920 kJ mol^{-1}, or 9.6 eV, which is close to the band gap of 8.8 eV calculated from the optical absorption edge in the ultra-violet spectral range (see Sect. 9.2 on optical absorption). The low activation energy portion at low temperatures was attributed to extrinsic electronic conductivity from ionization of impurities. The authors suggested that silicon as a donor atom was the most likely impurity resulting in the low temperature conductivity. The interpretation of extrinsic conduction in the low activation range agrees well with the results of several other studies of the electrical conductivity of alumina [35–38], which showed close to the same conductivity and activation energy at high temperatures, but a transition to the low activation energy regime at higher temperatures than 700°C, presumably because of more impurities in the samples in those studies.

The electrical conductivity of alumina parallel to the c-axis was found to be a factor of 3.3 higher than perpendicular to this axis [34].

Of special interest are some experimental results for the conductivity at temperatures from about 1,800°C to near the melting temperature of 2,054°C of alumina [39], which fall very close to an extrapolation of the data from [34] up to 1,300°C, with the same activation energy. Thus the intrinsic electrical conductivity σ in/ohm cm from 700°C to the melting point follows the equation:

$$\log \sigma = 7.92 - 24{,}200 / T \tag{9}$$

where T is in Kelvin.

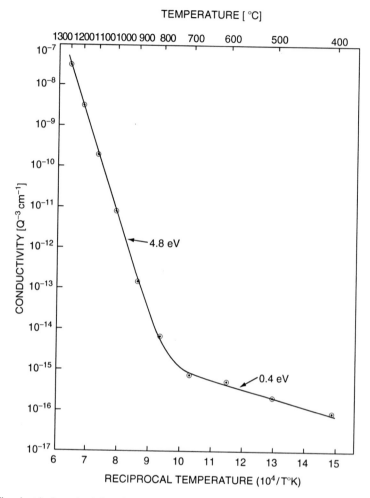

Fig. 3 The electrical conductivity of pure, dry sapphire along the *c*-axis. Points, measured values. From [34]

The electrolysis experiments of Ramirez et al. [40] show that when alumina contains some water (OH groups), the electrical conductivity results from the transport of hydrogen ions (actually hydronium ions, H_3O^+; see [41] for discussion).

The diffusion coefficient of H_3O^+ ions at 1,300°C calculated [41] from the experiments in [40] is $2.3(10)^{-9}\,cm^2\,s^{-1}$. This value is close to measured values of the diffusion coefficients of water in alumina [42]. Thus the mechanism of the diffusion of water in alumina is the transport of H_3O^+ ions, and these ions control the electrical conductivity when the water concentration is high enough.

To calculate the minimum concentration C of water in alumina that can contribute to the electrical conductivity, one can use the Einstein equation:

$$C = RT\sigma / Z^2F^2D \tag{10}$$

in which R is the gas constant, Z the ionic charge (valence), F the Faraday, and D the diffusion coefficient. The electrical conductivity at 1,300°C from [34] was 2.29

$\times 10^{-11}/\text{ohm cm}$. Thus with $D = 2.29(10)^{-9}\,\text{cm}^2\,\text{s}^{-1}$, a concentration of $1.13(10)^{-8}\,\text{mol}$ cm^{-3} of carriers results if one assumes that the conductivity in the samples in [34] results from H_3O^+ transport (which, of course, it does not); this concentration is $1.45(10)^{-7}$ carriers per Al atom in alumina. The concentration of H^+ in the alumina samples of [40] can be calculated from their highly sensitive infrared absorption measurements to be about $4.7(10)^{-7}$ per Al atom. Thus one can conclude that for H_3O^+ concentrations above about $10^{-8}\,\text{mol cm}^{-3}$ (3×10^{-7} ions per Al atom), there will be a contribution of these ions to the conductivity, whereas for lower H_3O^+ concentrations the conductivity will be mainly electronic.

The activation energy for water diffusion in alumina is about $220\,\text{kJ mol}^{-1}$ ($2.3\,\text{eV}$) from [42], so that many of the earlier results on electrical conductivity of alumina, for example, those summarized in [33], probably result from water transport at lower temperature; at higher temperatures, electronic conductivity will predominate, because of the high activation energy of intrinsic electronic conductivity. If the alumina is "dry" (H_3O^+ concentration below $10^{-8}\,\text{mol cm}^{-3}$) low activation energy extrinsic electronic conduction will be dominant at lower temperatures, resulting from donor and receptor impurities in the alumina.

7.2 Dielectric Properties

The dielectric constant of alumina is given in Table 19 as a function of temperature and crystal orientation. The dielectric constant increases slightly up to 500°C, and is quite dependent on orientation. Very low dielectric loss values for sapphire have been reported [2], but are questionable. With reasonable purity, loss tangents below 0.001 are likely. Actual values probably depend strongly on crystal purity.

7.3 Magnetic Properties

Alumina is diamagnetic with a susceptibility less than 10^{-6} [2].

8 Diffusion in Alumina

Experimental volume diffusion coefficients of substances in alumina are summarized in ref. 41. Values for the parameters D_0 and Q (activation energy) from the Arrhenius equation:

$$D = D_0 \exp(-Q / RT) \tag{11}$$

Table 19 Dielectric constant of sapphire as a function of temperature at frequencies from 10^3 to 10^9 Hz (from [2]). Orientation to c axis

Temp. (°C)	I	II
25	9.3	11.5
300	9.6	12.1
500	9.9	12.5

The fastest diffusing substance in alumina is hydrogen (H_2). Fast-diffusing cations are sodium, copper, silver, with hydroniums (H_3O^+) the fastest of these monovalent cations. Many other di- and trivalent cations have diffusion coefficients intermediate between these fast-diffusing ions and the slowest diffusers, the lattice elements aluminum and oxygen, which have about the same diffusion coefficients.

A number of experimenters have calculated diffusion coefficients D from "tails" on diffusion profiles in alumina, and attributed these D values to diffusion along dislocations, subboundaries, or grain boundaries. However, this attribution is doubtful in most cases, as discussed in [41]. In only two studies [43, 44] is it likely true diffusion along grain boundaries or dislocations was measured [41]. Mechanisms of diffusion in alumina are uncertain; a variety of charged defects have been suggested to control diffusion in alumina, but no interpretation is widely accepted because of discrepancies with experimental results. I have suggested that oxygen and aluminum diffusion in alumina results from transport of aluminum monoxide (AlO), and that AlO defects in the alumina structure are important in diffusion. These speculations have some support, but need more work to confirm them.

9   Chemical Properties

The decomposition of alumina at high temperatures can be deduced from its vapor pressure; see Sect. 6.3 and Table 15.

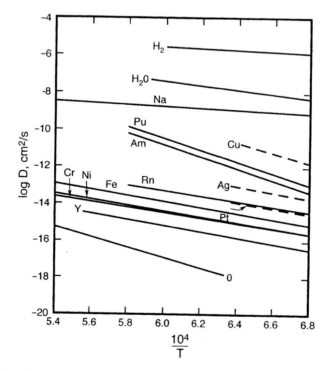

Fig. 4 Log diffusion coefficients vs. $10^4/T$ for selected substances diffusing in alumina. From [41]

9.1 Reactions with Metals

The chemical reactions of alumina with other substances can best be explored from the thermodynamic properties of these reactions. If the Gibbs free energy ΔG of the reaction at a particular temperature is negative, the reaction tends to take place, and if ΔG this energy is positive the reaction tends not to occur. These considerations are modified by concentrations (more properly, thermodynamic activities) of the components as expressed in an equilibrium constant of the reaction. See books on thermodynamics for more details, for example [45].

The relative Gibbs free energies of the reactions of metals with oxygen tell whether or not a particular metal will displace the aluminum in alumina. The reaction of aluminum with oxygen is

$$\frac{4}{3}Al + O_2 = \frac{2}{3}Al_2O_3 \qquad (12)$$

All of the oxidation reactions with metals are written with one mole of O_2 reacting for consistent comparison. This is the format for reactions plotted in an Ellingham diagram (see [46]). The Gibbs free energies of some of these oxidation reactions are given in Table 20, taken from [28, 29, 46–48]. If the Gibbs free energy of the reaction is higher than that of aluminum ($-845.6\,kJ\,mol^{-1}$ for reaction (12)), then this metal will react with alumina, displacing all the aluminum in any alumina in contact with the metal, either solid, liquid, or vapor. For example for yttrium:

$$\frac{4}{3}Y + O_2 = \frac{2}{3}Y_2O_3 \qquad (13)$$

the Gibbs free energy is $-1017\,kJ\,mol^{-1}$, so yttrium will displace aluminum from alumina:

$$2Y + Al_2O_3 = 2Al + Y_2O_3 \qquad (14)$$

The free energy change of reaction 14 can be deduced from the values for Eqs. (12) and (13) from Table 20 to be $-256\,kJ\,mol^{-1}$, showing the tendency for yttrium to displace aluminum. If the free energy shown in Table 20 is less than that for oxidation of aluminum, the metal will not react with alumina. Thus at 1,000°C, sodium and potassium vapors (1 atm) do not react with alumina, but 1 atm. of lithium vapor does. Liquid alkaline earths metals such as Mg, Ca, Sr, and Ba react with alumina at 1,000°C and displace aluminum metal. The relative tendency of reactions of solid and liquid metals with alumina does not change much with temperature. Of course at low temperatures (below about 500°C), the rates of reactions can be slow, even if the thermodynamics show a tendency to react. Gibbs free energies for other temperatures can be calculated from data in the thermodynamic tables in [28, 29, 46–48].

9.2 Reactions with Nonmetals

The halides Cl_2, Br_2, and I_2 do not react with alumina, but fluorine (F_2) does:

$$2Al_2O_3 + 6F_2 = 4AlF_3 + 3O_2 \qquad (15)$$

Table 20 Gibbs Free energies of reactions of metals with 1 atm. of oxygen at 1,000°C from [28, 29, 46–48]

Solid metals	$-\Delta G$ (kJ mol^{-1})	Liquid metals	$-\Delta G$
Y	1,017		
Zr	849	Ca	1,013
Ti	678	CE	962
Si	644	Mg	937
V	619	Ba	879
Mn	586	Li	870
Cr	544	Al	845.6
One atm. of metal vapor	$-\Delta G$ (kJ/mol)		
Na	444		
Zn	418		
K	326		

Metal + O_2 = oxide

At 1,000°C the Gibbs free energy for this reaction is about −2133 kJ mol^{-1}, showing a strong tendency to react.

Reactions of the gases H_2, H_2O, CO, and CO_2 with alumina can be deduced from the reactions:

$$Al_2O_3 + 3H_2 = 2Al + 3H_2O \tag{16}$$

$$Al_2O_3 + 3CO = 2Al + 3CO_2 \tag{17}$$

At equilibrium at 1,000°C the ratio of H_2/H_2O is about 10^{10} and CO/CO_2 about $2(10)^{10}$. It is impossible to reduce the water or carbon dioxide levels so low in any practical reaction process, so effectively H_2 and CO do not reduce Al_2O_3. Even at 2,000°C these ratios are about $3(10)^4$ and $2(10)^5$, which are difficult levels to maintain in practice. A possible reaction of alumina with carbon is

$$2Al_2O_3 + 6C = Al_4C_3 + 3CO_2 \tag{18}$$

However, the Gibbs free energy of this reaction is +1131 kJ mol^{-1} at 1,000°C and +504.6 kJ mol^{-1} at 2,000°C, so it will take place only at very low carbon dioxide concentrations.

Other chemical reactions of alumina can be examined with the thermodynamic data in [28–30, 47–48].

9.3 Reactions of Alumina in Aqueous Solutions

Alumina is amphoteric, which means that it dissolves in acidic and basic solutions, but not in neutral aqueous solutions. The solubility of alumina in solutions of pH from about 4–9 is low; at 25°C it is less than 10^{-7} mol l^{-1} at pH 6 [1]. Alternatively, alumina dissolves readily in strong acids (HCl, HNO_3, H_2SO_4) and strong bases (NaOH, KOH) at temperatures well above ambient (e.g., 90°C).

10   Optical Properties

10.1   Refractive Index

The optical properties of solids can be studied with the complex refractive index n^*:

$$n^* = n + ik \tag{19}$$

in which n is the real part of the refractive index and k is the imaginary part or absorption index. Values of n and k from $0.008731\,\mu m$ to $600\,\mu m$ (142.000–$0.00207\,eV$) are given to high accuracy and many wavelengths for alumina in [49]. In the wavelength range from $0.1454\,\mu m$ to $4.000\,\mu m$ (8.529–$0.31\,eV$), the value of k for highly pure alumina is less than 10^{-6} [49, 50], so the alumina is effectively transparent. Values of n for this wavelength range are given to four significant figures in Table 21. Values of n in the fifth or sixth significant figure are different for different investigations, probably because of different purities of samples and different measuring techniques and errors. A two-term equation for n (ordinary ray) in this wavelength range is [2]

$$n = 1.74453 + \frac{0.0101}{\lambda - 0.1598} \tag{20}$$

with λ the wavelength in micrometers. This equation gives n accurately to about four significant figures (Table 21). A more accurate three-term Sellmeier equation is

$$n^2 - 1 = \frac{A_1 \lambda^2}{\lambda^2 - \lambda_1^2} + \frac{A_2 \lambda^2}{\lambda^2 - \lambda_2^2} + \frac{A_3 \lambda^2}{\lambda^2 - \lambda_3^2} \tag{21}$$

with the constants A_i and λ_i given in Table 22.

10.2   Optical Absorption

The absorption limit of $8.73\,eV$ in Table 23 is close to the band gap of $8.8\,eV$ for alumina. At energies lower than $8.73\,eV$, trace impurities and defects in the alumina lead to absorption tails as described in [49]. Values of n and k at higher energies than $8.73\,eV$ are given in Table 23 to three or four significant figures. Values from different research groups can vary substantially [49]; those in Table 23 are from [50]. See [49] for n and k values at many more energies (wavelengths).

The values of k in Table 23 show a maximum at about $13\,eV$, which can be attributed to exitonic absorption [49]; other electronic processes in the ultra-violet spectral range are also described in [49].

Appreciable absorption begins in the infrared spectral range above a wavelength of $4.0\,\mu m$, as shown in Table 24; there are absorption peaks at $17.24\,\mu m$ ($580\,cm^{-1}$) and $22.73\,\mu m$ ($440\,cm^{-1}$), which result from lattice vibrations. For more details see [49].

The optical anisotropy of alumina results in slightly different n and k values for the ordinary and extraordinary ray, as shown in detail in [49] (see Table 22). This anisotropy is related to the hexagonal (rhombohedral) structure of the alumina.

Table 21 Refractive Index of sapphire at 25°C in the spectral range where $k < 10^{-6}$, from [49, 50]

Wavelength, λ, (μm)	Refractive index, n	Wavelength, λ (μm)	Refractive index, n
0.1464	2.231	1.367	1.749
0.1603	2.070	1.530	1.747
0.1802	1.947	1.709	1.743
0.200	1.913	1.960	1.738
0.220	1.878	2.153	1.734
0.240	1.854	2.325	1.731
0.260	1.838	3.244	1.704
0.280	1.824	3.507	1.695
0.300	1.815	3.707	1.687
0.330	1.804	4.000	1.675
0.361	1.794		
0.405	1.786		
0.436	1.781		
0.486	1.775		
0.546	1.771		
0.579	1.769		
0.644	1.765		
0.707	1.763		
0.852	1.759		
0.894	1.758		
1.014	1.755		

Table 22 Refractive index [16], constants at 20°C from 0.2 to 5.5 mm, from [49, 50]

Term	Ordinary ray	Extraordinary ray
$λ_1$	0.0726631	0.0740298
$λ_2$	0.1193242	0.1216529
$λ_3$	18.028251	20.072248
A_1	1.4313493	1.5039759
A_2	0.65054713	0.5506141
A_3	5.3414021	6.5927379

The refractive index of alumina increases slightly with increasing temperature T in the visible and near-visible spectral range. The value of the temperature coefficient dn/dT of the refractive index from wavelengths 0.4–0.8 μm is about $13(10)^{-6°}K^{-1}$ [49].

10.3   Color of Alumina

Without impurities alumina is colorless. However, addition of transition metal ions to alumina leads to spectacular colors, gem stones, and practical applications such as ruby lasers. Many aspects of color are discussed in detail in [51].

With the addition of about 1% of Cr_2O_3 to Al_2O_3 (replacement of one out of one hundred of the aluminum ions with chromium ions), alumina acquires a beautiful red color and is known as ruby, one of the most prized gem stones. The red color results from transitions of electrons between energy levels in the ruby, as described in [51], p. 8Iff. Ruby also shows a bright red fluorescence when it is illuminated with ultra-violet light (energy of 4–5 eV). Ruby also shows pleochoism (multicolors); in polarized light the color changes as the ruby crystal is rotated [51].

Table 23 Refractive index, n, and absorption index, k, of the ordinary ray for sapphire in the ultra-violet spectral range, at 25°C, from [49, 50]

Energy (eV)	Wavelength, λ, (μm)	n	k
142.000	0.008731	0.975	0.0130
120.000	0.010330	0.972	0.0177
100.206	0.01237	0.892	0.088
79.996	0.01550	0.929	0.130
60.001	0.02066	0.911	0.136
42.000	0.02952	0.859	0.098
30	0.04133	0.841	0.367
27.013	0.04590	0.605	0.317
24.113	0.05142	0.524	0.542
19.992	0.06202	0.800	0.983
18.008	0.06885	1.028	1.093
15.475	0.08012	1.142	1.125
13.582	0.09129	1.375	1.434
12.972	0.09558	2.581	1.573
12.361	0.1003	1.981	1.476
11.263	0.1101	2.400	1.131
10.292	0.1205	2.441	0.831
9.523	0.1302	2.559	0.686
8.852	0.1401	2.912	0.125
8.791	0.1410	2.841	0.037
8.760	0.1415	2.797	0.011
8.730	0.1420	2.753	0

Table 24 Refractive index, n, and absorption index, k, of the ordinary ray for sapphire in the infrared spectral range, at 25°C, from [49]

Wavelength, λ, (μm)	n	k
4.000	1.675	$1.30\,(10)^{-6}$
5.000	1.624	$3.76\,(10)^{-5}$
7.143	1.459	$3.60\,(10)^{-3}$
10.00	0.88	0.053
10.99	0.27	0.224
12.05	0.08	1.04
12.99	0.07	1.57
14.93	0.10	2.96
16.13	0.47	3.56
16.95	0.75	6.30
17.24	1.69	11.28
17.86	9.10	1.53
19.23	3.82	0.115
20.00	2.64	0.106
22.73	10.08	14.01
25.00	4.64	0.156
30.30	3.95	0.0145
33.33	3.69	0.00668
50.00	3.26	0.0117
60.13	3.19	0.0130
82.71	3.13	0.0085
100.0	3.12	0.0070
200.0	3.08	0.0035
333.3	3.07	0.0027
600.0	3.05	0.0015

The deep blue color or sapphire gems results from the addition of a few hundredths of one percent of iron and titanium impurities to alumina. The Fe^{2+} and Ti^{4+} ions substitute for aluminum in the sapphire, and when light of energy of 2.11 eV is shone on the sapphire, it is absorbed by the charge transfer reaction:

$$Fe^{2+} + Ti^{4+} = Fe^{3+} + Ti^{3+} \tag{22}$$

See [51], p. 140ff for a complete description of this process.

A variety of other colors are found in natural and synthetic alumina crystals [2, 51]. For example, an orange-brown color is produced by Cr^{4+} (padparadscha sapphire) [51]; different transition metal ions in different concentrations and oxidation states produce many colors.

11 Conclusion and Future Uses of Alumina

The properties of alumina listed in Sects. 4–10 show the unusual performance of pure alumina, leading to the variety of applications given in Table 1. Practical aluminas with impurities and defects have somewhat degraded properties, but often are superior to many other materials, and have a variety of specialized applications such as refractories, electronic components, and catalyst substrates. In [1] there are articles discussing the future of alumina. There will continue to be incremental improvements in processing methods and properties, leading to expansion of present applications. What really new areas of application of alumina are likely? These predictions are speculative, but the most promising new applications of alumina will probably be in electronic circuits, optical components, and biomaterials. Alumina fibers for composites and optics are attractive if they can be made pure, defect-free, and cheap. Because of its excellent properties other unsuspected applications of alumina will undoubtedly be developed.

Reference

1. L.D. Hart (ed.), *Alumina Chemicals*, The American Ceramic Society, Westerville, OH, 1990.
2. W.H. Gitzen, *Alumina as a Ceramic Material*, The American Ceramic Society, Westerville, OH, 1970.
3. Emphasizes defects and interfaces, especially grain boundaries, in alumina, *J. Am. Ceram. Soc.* 77, [2] (1994).
4. Emphasizes grain boundaries, grain growth, and diffusion in alumina, *J. Am. Ceram. Soc.* 86 [4] (2003).
5. P. Richet, J.A. Xu, and H.K. Mao, Quasi-hydrostatic compression of ruby to 500 kbar, *Phys. Chem. Min.* 16, 207–211 (1988).
6. A.P. Jephcoat, R.J. Hemley, H.K. Mao, and K.A. Goettel, X-ray diffraction of ruby (Al_2O_3, Cr^{3+}) to 175 GPa, *Physica B* 150, 115–121 (1988).
7. K.T. Thomson, R.M. Wentzcovitch, and M.S.T. Bukowinski, Polymorphs of alumina predicted by first principles, *Science* 274, 1880–1882 (1996).
8. A.H. Carim, G.S. Rohrer, N.R. Dando, S.Y. Tzeng, C.L. Rohrer, and A.J. Perrotta, Conversion of diaspore to corundum, *J. Am. Ceram. Soc.* 80, 2677–80 (1997).
9. R.L. Coble, Transparent Alumina and Method of Preparation, U.S. Patent 3,026,210, March 20 (1967).

10. F.J. Klug, S. Prochazka, and R.H. Doremus, Alumina-silica phase diagram in the mullite region, *J. Am. Ceram. Soc.* **70**, 750–759 (1987).
11. E.N. Bunting, Phase equilibrium in the system Cr3O3–Al2O3, *Bur. Stand. J. Res.* **6**, 947–949 (1931).
12. E.M. Levin, C.R. Robbins, and H.F. *McMurdie, Phase Diagrams for Ceramists*, Vol. I–XIII, The American Ceramic Society, 1964–2002.
13. D.M. Roy and R.E. Barks, Subsolidus phase equilibrium in Al2O3–Cr2O3, *Nature*, **235**, 118–119 (1972).
14. C. Greskovich and J.A. Brewer, Solubility of magnesia in polycrystalline alumina at high temperatures, *J. Am. Ceram. Soc.* **84**, 420–425 (2001).
15. J.H. Gieske and G.R. Barsch, Pressure dependence of the elastic constants of single crystalline aluminum oxide, *Phys. Status Solidi* **29**, 121–131 (1967).
16. A. Kelly, *Strong Solids*, Chap. 1, Oxford University Press, London, 1973.
17. S.M. Wiederhorn in Mechanical and Thermal Properties of Ceramics, J.B. Wachtman, (ed.), NBS Special Publications 303, U.S. GPO, Washington, D.C., 1969, p. 217.
18. R.H. Doremus, Cracks and energy-criteria for brittle fracture, *J. Appl. Phys.* **47**, 1833–1836 (1976).
19. R.H. Doremus, Fracture statistics: A comparison of the normal, Weibull and type I extreme value distributions, *J. Appl. Phys.* **54**, 193–201 (1983).
20. S.C. Carniglia, Reexamination of experimental strength – vs. – grain – size data for ceramics, *J. Am. Ceram. Soc.* **55**, 243 (1972).
21. J.E. Burke, R.H. Doremus, W.B. Hillig, and A.M. Turkalo, Static Fatigue in Glasses and Alumina, in Materials Sci. Res. Vol. 5, W.W. Kriegel (ed.), Plenum Press, New York, 1971, pp. 435–444.
22. N.W. Thibault and H.Z. Nyquist, The measured Knoop hardness of hard substances and factors affecting its determination, *Trans. Am. Soc. Metals* **38**, 271–330 (1947).
23. W.D. Kingery, H.K. Bowen, and D.R. Uhlmann, *Introduction to Ceramics*, Wiley, New York (1976).
24. W.R. Cannon and T.G. Langdon, Creep of Ceramics, *J. Mater. Sci.* **18**, 1–50 (1983); **23**, 1–20 (1988).
25. A.H. Hynes and R.H. Doremus, Theories of creep in ceramics, *Crit. Rev. Solid State Mater. Sci.* **21**, 1–59 (1996).
26. M.L. Kronberg, Plastic deformation of single crystals of sapphire: Basal slip and twinning, *Aeta Met.* **5**, 507–529 (1957).
27. J.D. Snow and A.H. Heuer, Slip systems in Al2O3, *J. Am. Ceram. Soc.* **56**, 153–157 (1973).
28. M.W. Chase, NIST – JAVAF Thermochemical Tables, *J. Phys. Chem. Ref. Data, Monograph* **9**.
29. R.H. Tamoreaux, D.L. Hildenbrand and L. Brewer, High temperature vaporization behavior of oxides, *J. Phys. Chem. Ref. Data* **16**, 412 (1987).
30. L. Brewer and A.W. Searcy, Gaseous species of the Al–Al$_2$O$_3$ system, *J. Am. Chem. Soc.* **73**, 5308–5314 (1951).
31. R.V. Gains, H.C.W. Skinner, E.E. Foord, B. Mason, and A. Rosenzweig, *Dana's New Mineralogy*, Wiley, New York, 1977, p. 214.
32. E. Schreiber and O.L. Anderson, Pressure derivatives of the sound velocities of polycrystalline alumina, *J. Am. Ceram. Soc.* **49**, 184–190 (1966).
33. J. Pappis and W.D. Kingery, Electrical properties of single and polycrystalline alumina at high temperatures, *J. Am. Ceram. Soc.* **44**, 459 (1961).
34. F.G. Will, H.G. Lorenzi, and K.H. Janora, *J. Am. Ceram. Soc.* **75**, 295–304, 2790–2791 (1992).
35. O.T. Özkan and A.J. Moulson, The electrical conductivity of single crystal and polycrystalline aluminum oxide, *British J. Appl. Phys.* **3**, 983 (1970).
36. H.P.R. Frederike and W.R. Hosler, High temperature electrical conductivity of aluminum oxide, *Mater. Sci. Res.* **9**, 233 (1973).
37. K. Kituzawa and R.L. Coble, Electrical conduction in single crystal and polycrystalline Al2O3 at high temperature, *J. Am. Ceram. Soc.* **57**, 245 (1979).
38. H.M. Kizilyalli and P.R. Mason, DC and AC electrical conduction in single crystal alumina, *Phys. Status Solidi* **36**, 499 (1976).
39. E.E. Shpilrain, D.N. Kagan, L.S. Barkhatov, and L.I. Zhmakin, The electrical conductivity of alumina near the melting point, *High Temperatures – High Pressures* **8**, 177 (1976).
40. R. Ramirez, R. Gonzalez, J. Colera, and Y. Chen, Electric-field-enhanced diffusion of deuterons and protons in α-Al2O3 crystals, *Phys. Rev. B* **55**, 237–242 (1997).
41. R.H. Doremus, Diffusion in Alumina, *J. Appl. Phys.* **101**, 101301 (2006).

42. J.D. Fowler, D. Chandra, T.S. Elleman, A.W. Payne, and K. Verghese, Tritium diffusion in Al2O3 and BeO, *J. Am. Ceram. Soc.* **60**, 55 (1977).
43. V.S. Stubican and J.W. Orenbach, Influence of anisotropy and doping on grain-boundary diffusion in oxide systems, *Solid State Ionics* **12**, 375 (1984).
44. X. Tang, and K.P.D. Lagerof, and A.H. Heuer, Determination of pipe diffusion coefficients in undoped and magnesia doped sapphire (α-Al2O3); a study based on annihilation of dislocation dipoles, *J. Am. Ceram. Soc.* **86**, 560–565 (2003).
45. D.R. Gaskell, *Introduction to the Thermodynamics of Materials*, 4th edn., Taylor and Francis, New York, 2003.
46. L.S. Darken and N.W. Gurry, *Physical Chemistry of Metals*, McGraw-Hill, New York, 1953.
47. A. Paul, *Chemistry of Glasses*, Chapman and Hall, London, 1982, p. 157.
48. R.H. Doremus, *Diffusion of Reactive Molecules in Solids and Melts*, Wiley, New York, 2002, pp. 71, 191.
49. M.E. Thomas and W.J. Tropf in *Handbook of Optical Constants of Solids III*, E.D. Palik (ed.), Academic Press, New York, 1998, p. 653.
50. L.H. Malitson and M.J. Dodge, Refractive index and birefringence of synthetic sapphire, *J. Opt. Soc. Am.* **62**, 1405 (1972).
51. K. Nassau, *The Physics and Chemistry of Color*, Wiley, New York, 1983.

Chapter 2
Mullite

David J. Duval, Subhash H. Risbud, and James F. Shackelford

Abstract Mullite is the only stable intermediate phase in the alumina–silica system at atmospheric pressure. Although this solid solution phase is commonly found in human-made ceramics, only rarely does it occur as a natural mineral. Yet mullite is a major component of aluminosilicate ceramics and has been found in refractories and pottery dating back millennia. As the understanding of mullite matures, new uses are being found for this ancient material in the areas of electronics and optics, as well as in high temperature structural products. Many of its high temperature properties are superior to those of most other metal oxide compounds, including alumina. The chemical formula for mullite is deceptively simple: $3Al_2O_3 \cdot 2SiO_2$. However, the phase stability, crystallography, and stoichiometry of this material remain controversial. For this reason, research and development of mullite is presented in an historical perspective that may prove useful to engineers and scientists who encounter this material under nonequilibrium conditions in their work. Emphasis is placed on reviewing studies where the primary goal was to create single-phase mullite monoliths with near theoretical density.

1 Introduction

Mullite is a solid solution phase of alumina and silica commonly found in ceramics. Only rarely does mullite occur as a natural mineral. According to introductory remarks made by Schneider and MacKenzie at the conference "Mullite 2000"[1], the geologists Anderson, Wilson, and Tait of the Scottish Branch of His Majesty's Geological Survey discovered the mineral mullite less than a century ago. The trio was collecting mineral specimens from ancient lava flows on the island of Mull off the west coast of Scotland when they chanced upon the first known natural deposit of this ceramic material. The specimens were initially identified as sillimanite, but later classified as mullite.

Being the only stable intermediate phase in the Al_2O_3–SiO_2 system at atmospheric pressure, mullite is one of the most important ceramic materials. Mullite has been fabricated into transparent, translucent, and opaque bulk forms. These materials may have optical and electronic device applications. Mullite's temperature stability and

J.F. Shackelford and R.H. Doremus (eds.), *Ceramic and Glass Materials:*
Structure, Properties and Processing.
© Springer 2008

refractory nature are superior to corundum's in certain high-temperature structural applications. Another characteristic of this aluminosilicate is its temperature-stable defect structure, which may indicate a potential use in fuel cell electrolytes.

In this chapter, developments in the understanding of mullite over the last few decades are reviewed. A discussion of crystal structures and phase stability is presented to provide the reader with an overview of certain characteristics of this material. The next part of this chapter examines the effect of process chemistry on the synthesis and microstructure of mullite. The role of various synthetic methods that are used to modify mullite formation will be discussed, followed by a compilation of selected materials properties.

2 Crystal Structure

The X-ray diffraction pattern of mullite is very similar to that of sillimanite. Sillimanite is a commonly occurring aluminosilicate mineral stable at high pressures with the chemical formula $Al_4Si_2O_{10}$, a 1:1 ratio of silica to alumina.

Roughly speaking, the sillimanite and mullite structures consist of chains of distorted edge-sharing Al–O octahedra at the corners and center of each unit cell running parallel to the c-axis. The chains are cross-linked by Si–O and Al–O corner-sharing tetrahedra [2]. Mullite is a solid solution compound with stoichiometries ranging from relatively silica-rich $3Al_2O_3 \cdot 2SiO_2$ (3:2 mullite) to alumina-rich $2Al_2O_3 \cdot SiO_2$ (2:1 mullite). The structure of mullite is summarized in Table 1. Some authors use the Al/Si ionic ratio when referring to mullite stoichiometry. In this case, 3:2 mullite would have an aluminum/silicon ionic ratio of 3:1. To avoid further confusion and follow the convention most commonly used in the literature, mullite stoichiometry will be based on the alumina/silica molecular ratio. The chemical formula for mullite is often given by $Al_2(Al_{2+2x}Si_{2-2x})O_{10-x}$, where $x = 0$ corresponds to sillimanite, $x = 0.25$ corresponds

Table 1 Wyckoff positions and coordinates of atom sites for the orthorhombic mullite structure with space group Pbam (No. 55)

Lattice parameters	$a = 0.75499(3)$ nm		$b = 0.76883(3)$ nm			$c = 0288379(9)$ nm	
Atom	Al_2	$[Al_2Si_{2-2x}]$	Al_{2x}	O_{2-3x}	O_{2x}	O_4	O_4
Wyckoff position	2a	4h	4h	2d	4h	4h	4g
Coordinate							
x	0	0.1474(6)	0.268(3)	0.5	0.451(5)	0.3566(6)	0.1263(9)
y	0	0.3410(6)	0.207(2)	0.0	0.048(5)	0.4201(6)	0.2216(8)
z	0	0.5	0.5	0.5	0.5	0.5	0.0
Thermal parameter (β)	0.5(1)	0.3(1)	1.2(8)	0.8(1)	0.8(1)	0.8(1)	0.8(1)
Occupancy							
O	1	0.5	0.166(7)	0.5	0.166(7)	1	1
Al		0.334(7)					
Si							

The chemical formula is $Al_2(Al_{2+2x}Si_{2-2x})O_{10-x}$, where $x = 0.33$ and the calculated density is 3.16 g cm⁻³. From [57]

to 3:2 mullite, and $x = 0.4$ corresponds to 2:1 mullite. Diffusion studies [3] have shown that the following chemical formula is more appropriate even though it is not commonly seen in the literature:

$$\text{Al}^{\text{VI}}_{\left(\frac{4}{3}+\frac{14}{3}x\right)}\left[\text{Al}^{\text{IV}}_{\frac{8}{3}}\text{Si}_2\right]_{(1-x)}\text{O}_{10-x}\square_x \tag{1}$$

The symbol $\square$ denotes an oxygen vacancy. The superscripts VI and IV indicate octahedral and tetrahedral coordination sites, respectively.

With increasing alumina content, Si^{4+} is replaced by Al^{3+} and anion (oxygen) vacancies are created to maintain charge neutrality. Accommodating the structural defects causes significant distortions of the aluminum and silicon polyhedra. In mullite (as opposed to sillimanite), there are three (as opposed to four) tetrahedral "chains" in the unit cell, with a somewhat random distribution for silica and alumina tetrahedra [4]. This results in the necessity for distorted alumina tetrahedra to be arranged in an oxygen-deficient tricluster (three tetrahedra sharing single corner-bridging oxygen). These clusters constitute a distinctive element of mullite's crystal structure [2,5].

Unlike sillimanite, X-ray diffraction patterns of mullite exhibit significant diffuse scattering and possible superlattice reflections. Authors have proposed various models to account for mullite's anomalous scattering using superlattice refinement, atomic site occupancy factor calculation, and correlated vacancy mapping [2,6,7]. Most work suggests that defects tend to cluster or correlate with short-range order along specific crystallographic directions. Lower alumina concentrations result in less directional correlation of oxygen vacancies or more random vacancy distributions. According to Freimann and Rahman [7], oxygen vacancies tend to correlate parallel with the lattice parameter a, and to a lesser extent with b. The authors suggest their correlation results could be used to interpret thermal expansion behavior of mullites. As a practical matter, the lattice parameter a correlates linearly with

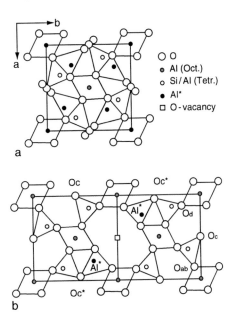

Fig. 1 Structure of mullite. (**a**) Average structure and (**b**) atomic displacements around an oxygen vacancy. From [7]

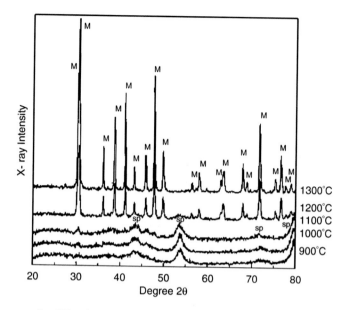

Fig. 2 X-ray powder diffraction patterns showing the crystallization of mullite from amorphous precursors as a function of temperature. M denotes mullite peaks, and Sp markers denote the intermediate γ-Al$_2$O$_3$ spinel peaks. From [8]

Al$_2$O$_3$ content. Figure 1 depicts the mullite unit. Atom positions for an intermediate composition of mullite, Al$_2$(Al$_{2+2x}$Si$_{2-2x}$)O$_{10-x}$, where $x = 0.33$ are provided in Table 1. X-ray powder diffraction patterns demonstrating mullite crystallization from amorphous precursors are shown in Fig. 2 [8].

It should be noted that there is no convincing evidence of mullite formation in regions of the phase diagram with compositions between 3:2 mullite and sillimanite. In other words, the chemical formula for mullite cannot accommodate x values such that $0<x<0.25$. Although the presence of a cubic spinel with the stoichiometry and structure similar to that of 2:1 mullite had been reported [9,10], its existence is likely of academic rather than practical significance. What was originally reported as a tetragonal phase of 3:1 mullite [11] formed by rapid quenching of the melt could be attributed to severe microtwinning of the usual orthorhombic structure [12]. On the other hand, workers have recently reported mullite phases with Al$_2$O$_3$/SiO$_2$ ratios up to and greater than 9:1 [13–15]. These specialty compounds are potentially useful in specific refractory applications due to their high Al$_2$O$_3$ content. Unfortunately, it has proved difficult to produce these ultra-high alumina mullites in sufficient quantity and purity. Further research is required before practical applications for these materials can be envisioned.

3 Phase Stability

An historical perspective may prove useful to engineers or scientists who encounter mullite during the course of their work: The earliest interpretations of the material's behavior may reflect the result of nonequilibrium conditions that often occur in production or experimental situations.

Mullite-based ceramics have been widely used as refractories and in pottery for millennia. Although the technology of mullite is becoming more mature, there are still questions concerning its melting behavior and the shape of mullite phase boundaries in the Al_2O_3–SiO_2 phase diagram. In 1924, Bowen and Grieg [16] published the first phase diagram to include mullite as a stable phase, but did not indicate a solid solution range. The phase $3Al_2O_3 \cdot 2SiO_2$ was reported to melt incongruently at 1,810°C. Specimens were prepared from mechanical mixtures of alumina and silica melted and quenched in air. Shears and Archibald [17] reported the presence of a solid solution range from $3Al_2O_3 \cdot 2SiO_2$ (3:2 mullite) to $2Al_2O_3 \cdot SiO_2$ (2:1 mullite) in 1954. Their phase diagram depicted a mullite solidus shifting to higher alumina concentrations at temperatures above the silica–mullite eutectic temperature.

In 1958, Toropov and Galakhov [18] presented a phase diagram where mullite was shown to melt congruently at 1,850°C. Aramaki and Roy [19] published a phase diagram in 1962 corroborating a congruent melting point for mullite at 1,850°C. Their specimens were prepared from gels for subsolidus heat treatments, while mechanical mixtures of α-Al_2O_3 and silica glass were prepared for heat treatments above the solidus temperature. Specimens were encapsulated to inhibit silica volatilization. A silica–mullite eutectic temperature of 1,595°C and a mullite–alumina eutectic temperature of 1,840°C were reported. No shift in the mullite solidus phase boundary with temperature was reported in either of these publications.

Over a decade later, Aksay and Pask [20] presented a different phase diagram depicting incongruent melting for mullite at 1,828°C. Specimens, in the form of diffusion couples between sapphire and aluminosilicate glass, were also encapsulated to inhibit volatilization. Many authors suggest that nucleation and growth of mullite occurs within an amorphous alumina-rich siliceous phase located between the silica and alumina particles [21–24]. On the other hand, Davis and Pask [25] and later Aksay and Pask observed coherent mullite growth on sapphire in a temperature range from about 1,600 to below 1,800°C, indicating interdiffusion of aluminum and silicon ions through the mullite [20]. Risbud and Pask [26] later modified the diagram to incorporate metastable phase regions. They showed a stable silica–mullite eutectic temperature of 1,587°C. An immiscibility dome with a spinodal region was reported between approximately 7 and 55 mol% Al_2O_3. The dome has a central composition of about 35 mol% Al_2O_3, and complete miscibility occurs near 1,550°C (temperatures below the silica–mullite eutectic temperature). A stable mullite–alumina peritectic was reported at 1,828°C. However, a "metastable" incongruent melting point for mullite was reported at 1,890°C. The "metastable" mullite compositions were shifted toward higher alumina concentration. To account for the metastability, the authors suggested there could be a barrier for alumina precipitation in both melt and mullite, and that mullite could be superheated. Figure 3 portrays this phase diagram showing regions of metastability [27].

In 1987, Klug et al. published their SiO_2–Al_2O_3 phase diagram [28]. They reported incongruent melting for mullite at 1,890°C, and shifting of both boundaries of the mullite solid solution region toward higher alumina content (2:1 mullite) at temperatures above the eutectic point of 1,587°C. This phase diagram appears to reconcile most of the phenomena observed by other workers on the SiO_2–Al_2O_3 system. Seemingly irreconcilable observations involving phase stability of similarly prepared specimens have been attributed convincingly to nonequilibrium conditions and/or silica volatilization. This phase diagram [28] is shown in Fig. 4.

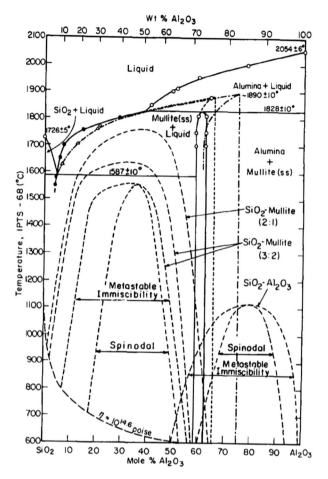

Fig. 3 The system Al$_2$O$_3$–SiO$_2$ showing metastable regions. The gaps shown with spinodal regions are considered the most probable thermodynamically. From [27]

The 2:1 mullite appears to be only metastable at room temperature [28], and very high temperature use or cycling might cause some alumina to precipitate. However, Pask [29] suggested that discrepancies in the reported behavior of mullite are attributable to the presence or absence of α-Al$_2$O$_3$ in the starting materials. Engineers or scientists are cautioned to use the appropriate phase diagram consistent with their experimental methods and conditions. It should also be noted that at tectonic pressures, SiO$_2$ will exsolve from mullite leaving a compound with a stoichiometry Al$_2$O$_3$·SiO$_2$. Depending on temperature and pressure, the compound will be sillimanite, kyanite, or andalusite.

4 Processing and Applications

As mentioned in the previous section, the formation, phase purity, and morphology of mullite depend upon precursor materials and processing history. Mullite was first identified as the product of heating kaolinitic clays, resulting in a compound with an

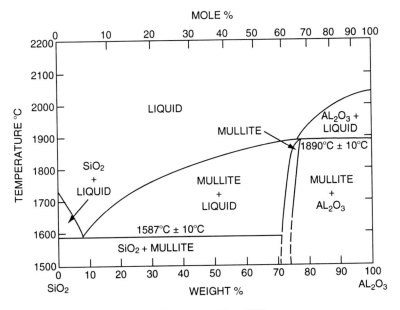

Fig. 4 Phase diagram for the alumina–silica system. From [28]

approximate alumina-to-silica molar ratio of 3:2. The order of reaction proceeds as follows [30]:

$Al_2Si_2O_5(OH)_4$	$\xrightarrow{450°C}$	$2(Al_2O_3·2SiO_2) + 2H_2O$
Kaolinite		Metakaolin
$2(Al_2O_3·2SiO_2)$	$\xrightarrow{925°C}$	$2Al_2O_3·3SiO_2 + SiO_2$
Metakaolin		Silicon spinel
$2Al_2O_3.3SiO_2$	$\xrightarrow{1,100°C}$	$2(Al_2O_3.SiO_2) + SiO_2$
Silicon spinel		Pseudomullite
$3(Al_2O_3.SiO_2)$	$\xrightarrow{1,400°C}$	$3Al_2O_3.2SiO_2 + SiO_2$
Pseudomullite		Mullite + cristobalite

Excess corundum may be added, and the system heated at higher temperatures to minimize free SiO_2. Toward this end, Goski and Caley [31] suspended grains of the mineral kyanite (a high-pressure form of $Al_2O_3·SiO_2$) with submicron alumina in water to provide intimate mixing of these mullite precursors. The alumina–kyanite suspension was slip cast to form a green body that was reaction-sintered to form an alumina–mullite composite. According to phase diagrams, a silica-rich glassy phase in 3:2 mullite is predicted when sintered at temperatures higher than the eutectic (1,587°C). Many common 3:2 mullite products are sintered between 1,600 and 1,700°C and may contain a glassy phase in the microstructure.

High-purity glass-free mullite monoliths have been obtained by at least three traditional methods:

1. Starting materials with alumina contents near the stoichiometry of 2:1 mullite may be completely melted above 1,960°C and then cooled to about 1,890°C without crystallizing. At the latter temperature (in the shifted solid solution region), infrared-transparent mullite single crystals could be grown by the Czochralski method [32].
2. Pask [29] reports that mullites with higher molar ratios of alumina to silica (i.e., >3:1) have been prepared by homogenous melting of the constituents above the liquids and subsequent quenching. As a note, mullites prepared by fusion are generally weaker than those produced by sintering [33].
3. Mullite powders obtained by various methods can first be crystallized near 1,200°C, and then sintered at temperatures below the eutectic. Highly pure mullite and mullite composites have been obtained by hot pressing below 1,300°C with this method [34].

When processed close to or above the eutectic temperature (~1,590°C), mullite with bulk compositions of less than 72 wt% Al_2O_3 (3:2 mullite) exhibits a microstructure of elongated grains that is believed to be promoted by the presence of a glassy second phase. For Al_2O_3 concentrations greater than 72 wt% Al_2O_3, the amount of glassy phase is less and the initially formed mullite grains are smaller and more equiaxial. Further heat treatment results in rapid grain growth driven by a decrease of the high grain boundary area associated with the fine grains in the initial system. This leads to fast growth of the grains along the c-axis and a higher aspect ratio for the overall grains. After this rapid decrease in the driving force, the grains grow more slowly and the overall decrease in the free energy of the system dictates the development of a more equiaxial microstructure [35]

An interesting approach in making mullite powders has been via combustion synthesis [36]. An aqueous heterogeneous redox mixture containing aluminum nitrate, silica fume (soot), and urea in the appropriate mole ratio is mixed together. When rapidly heated to 500°C, the mixture boils, foams, and can be ignited with a flame. The process yields weakly crystalline mullite powder in less than 5 min. Fully crystalline mullite can be obtained by incorporating an extra amount of oxidizer, such as ammonium perchlorate in the solution.

Recent work on mullite synthesis has focused on variations of sol–gel methods, which allow control of the local distribution and homogeneity of the precursor chemistry. The microstructure of a sol–gel derived mullite is shown in Fig. 5. Along with an understanding of kinetics, sol–gel methods look promising for use in the manufacture of bulk materials, thin films, or fibers of mullite with almost any specified phase purity, phase distribution, and grain morphology.

Three categories of gels are usually made [37]. Single-phase (type I) mullite precursor gels have near atomic level homogeneous mixing. The precursors transform into an alumina-rich mullite at about 980°C in the same way as rapidly quenched aluminosilicate glasses. These are formed from the simultaneous hydrolysis of the aluminum and silicon sources. Type I xerogels, for example, can be synthesized from tetraethylorthosilicate (TEOS) or tetramethylorthosilicate (TMOS) and aluminum nitrate nonahydrate [38]. Diphasic (type II) gels comprised two sols with mixing on the nanometer level. These gels, after drying, consist of boehmite and noncrystalline SiO_2, which at ~350°C transform to γ-Al_2O_3 and noncrystalline SiO_2. An example of

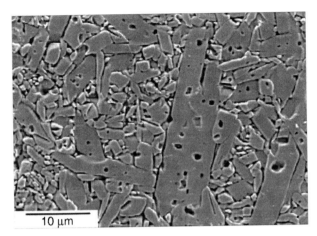

Fig. 5 Scanning electron micrograph of 3:2 mullite. Specimen was sintered at 1,700°C, hot isostatically pressed at 1,600°C, and thermally etched. From [54]

a type II gel would be a mixture of boehmite with a TEOS or TMOS sol [22]. Type III diphasic gels contain precursors that are noncrystalline up to 980°C and then form γ-Al_2O_3 and noncrystalline SiO_2.

Subsequent heat treatments of the three types of gels result in very different microstructures even if the alumina–silica molecular ratios are identical. Mullite conversion from powders or diphasic gels tends to be diffusion rate controlled. In the case of monophasic gels, conversion from the amorphous to crystalline phase appears to be nucleation rate dependent [39]. Such nucleation rate dependence would seem to indicate that it would be difficult to obtain very fine-grained mullite monoliths. However, some researchers have been successful in producing such monoliths. Monophasic xerogels prepared by slow hydrolysis (4–6 months) of hexane solutions of aluminum sec-butoxide and TMOS have been used to make optically clear mullite monoliths. The gel was heated in the range of ~1,000–1,400°C to form a completely dense crystalline material with glass-like mechanical properties (brittle and conchoidal fractures, rapid crack propagation, and no clear evidence of intergranular fracture) [40].

Seeding sol–gel precursors with nucleation sites for growth appears to be a method of making fine-grained monolithic optically transparent materials. Initially upon heating, gels formed by mixing a colloidal boehmite–silica sol with a polymeric aluminum nitrate–TEOS sol (a hybrid type I and type II gel) tend to crystallize, forming mullite seed crystals. Homoepitactic nucleation during continued heat treatment results in mullite monoliths. The introduction of the polymeric gel resulted in an increase in apparent nucleation frequency by a factor of 1,000 at 1,375°C, and a reduction in high-temperature grain size from 1.4 to 0.4 μm at 1,550°C, with little or no intragranular porosity [41].

MacKenzie et al. [42] prepared type I gels to determine the role of preheat treatment temperature on subsequent mullite microstructure. They found that an optimal preheat temperature of about 250–350°C for a long period of time resulted in an optimal concentration of mullite in the final product. Concurrently, there was an increase in the ^{27}Al nuclear magnetic resonance spectrum at about 30 ppm. The 30 ppm Al signal is often attributed to penta-coordinated Al, which may be located in the mullite precursor gels at the interface between Si-rich and Al-rich microdomains. MacKenzie et al. attribute this Al signal to the distorted tetrahedral Al environment in the region of O-deficient triclusters. They noted that the signal becomes increasingly strong just prior to mullitization. It was also noted that organic residues and hydroxyl groups were present up to 900°C. According to the analysis, the presence of these groups in the system at high temperatures could influence the structural evolution of the gel by providing a locally reducing and/or humid atmosphere that could facilitate tricluster formation. These sites could influence subsequent mullite formation because they form an essential element of the mullite structure. In terms of the nature of the triclusters, Schmueker and Schneider [5] proposed that the triclusters of tetrahedra may compensate the excess negative charge in the network caused by $Si^{+4}–Al^{3+}$ substitution. Na^+ doped into aluminosilicate gels can also compensate for the $Si^{4+}–Al^{3+}$ substitution. For this system, the formation of triclusters was no longer required, and a significant drop in the 30 ppm Al peak was observed.

Transparent mullite may have optical applications. With a scattering loss of less than $0.01\,cm^{-1}$, it could be an excellent candidate for use in transparent windows in the mid-infrared range (3–5 μm wavelength). Furthermore, when mullite glass ceramics were formed with Cr^{3+} additions, significant differences in the luminescence spectra between the glassy phase and crystalline mullite were observed [43] Cr^{3+} was shown to reside in the mullite crystalline phase. The luminescence quantum efficiency increased from less than 1% to about 30% by the crystallization process. Further research is needed to establish mullite as a candidate for high-energy laser applications.

5   Selected Materials Properties

The availability of fine, pure mullite powders and novel processing routes have made it possible to obtain dense polycrystalline mullite with higher deformation resistance and hardness at higher temperatures than most other ceramics, including alumina [44,45]. Mullite has good chemical stability and a stable temperature-independent oxygen vacancy structure up to the melting point [46], making mullite particularly creep-resistant. It should be noted that the majority of studies on high temperature mechanical properties of mullite have concentrated on measurements of strength or the creep deformation under testing conditions of four point bending or compression under static loading [47,48]. These testing procedures are useful as an initial evaluation of failure strength or creep resistance but the complexity of the stress makes it difficult to interpret the effect of the material variables on the creep mechanisms [49]. Nevertheless, to cite one representative study, creep may occur by a diffusional mechanism for grain sizes <1.5 μm with stresses of less than 100 MPa at temperatures between 1,365 and 1,480°C. High activation energy of 810 kJ mol^{-1} was determined for this process. Larger grain sizes and higher stresses indicate creep occurs by slow crack growth [48]. Selected mechanical properties are provided in Table 2. In general, creep resistance increases with sintering temperature, while flexural strength decreases [50].

With a low thermal conductivity of 0.06 W cm^{-1} K^{-1} and a low thermal expansion coefficient $\alpha \sim 4.5 \times 10^{-6}°C^{-1}$, mullite is useful for many refractory applications [49]. According to Schneider, most mullites display low and nonlinear thermal expansions below, but larger and linear expansion above, ~300°C. The volume thermal expansion

Table 2 Values of fracture toughness (K_{Ic}), fracture strength (σ_f), flexural strength, and microhardness for 3:2 mullite at different temperatures

T (°C)	K_{Ic} (MPa m$^{1/2}$)	σ_f (MPa)	Flexural strength (MPa)	Microhardness (GPa)
22	2.5 ± 0.5[a]			15[b]
1000				10[b]
1200	3.6 ± 0.1	260 ± 15	500[c]	
1300	3.5 ± 0.2	200 ± 20		
1400	3.3 ± 0.2	120 ± 25	360[c]	

From [49] (specimens had apparent density of 2.948 Mg m^{-3} and grain size of 4.0 μm)
[a] Value from [58]
[b] Values from [45]
[c] Values mentioned in [8]

decreases with alumina content, and the anisotropy of thermal expansion is reduced simultaneously [51].

Given that mullite is a defect structure, one would expect high ionic conductivity. Rommerskirchen et al. have found that mullite has ionic conductivity superior to that of the usual CaO-stabilized ZrO_2 solid electrolytes at temperatures from 1,400 to 1,600°C [52]. The oxygen self diffusion coefficient in the range $1,100 < T < 1,300$°C for a single crystal of 3:2 mullite has been given by [53]:

$$D_{ox} = 1.32 \times 10^{-2} \exp[-397 \, kJ / RT] cm^2 \, s^{-1} \tag{2}$$

Grain boundary diffusion coefficients are about five orders of magnitude higher than volume diffusion in the same temperature range. The activation energy for grain boundary diffusion [54] is 363 ± 25 kJ mol^{-1} – a remarkably similar value compared with that of volume diffusion.

The activation energy for silicon diffusion during the formation of mullite from fused couples at $1,600 < T < 1,800$°C [55] is in the range of $730 < \Delta H_{si}^{4+} < 780$ kJ mol^{-1}. There is support for the idea that Al^{3+} diffusion coefficients are much higher than those of silicon at temperatures above the mullite–silica eutectic [56].

References

1. H. Schneider and K. MacKenzie, *J. Eur. Ceram. Soc.* **21**, iii (2001).
2. M. Tokonami, Y. Nakajima, and N. Morimoto, The diffraction aspect and a structural model of mullite, $Al(Al_{1+2x}Si_{1-2x})O_{5-x}$, *Acta Cryst.* **A36**, 270–276 (1980).
3. J. L. Holm, On the energetics of the mullite solid-solution formation in the system Al_2O_3–SiO_2, *J. Mat. Sci. Lett.* **21**, 1551–1553 (2002).
4. W.M. Kriven, M.H. Jilavi, D. Zhu, J.K.R. Weber, B. Cho, J. Felten, and P. C. Nordine, Synthesis and microstructure of mullite fibers grown from deeply undercooled melts, in *Ceramic Microstructures: Control at the Atomic Level*, A. P. Tomsia and A. M. Glaeser (eds.), Plenum, New York, NY, (1998) pp. 169–176.
5. M. Schmuecker and H. Schneider, Structural development of single phase (type I) mullite gels, *J. Sol–Gel Sci. Tech.* **15**, 191–199 (1999).
6. R.X. Fischer, H. Schneider, and M. Schmuecker, Crystal structure of Al-rich mullite, *Am. Mineral.*, **79** (9–10), 983–990 (1994).
7. S. Freimann and S. Rahman, Refinement of the real structures of 2:1 and 3:2 mullite, *J. Eur. Ceram. Soc.* **21**, 2453–2461 (2001).

8. D.J. Cassidy, J.L. Woolfrey, and J.R. Bartlett, The effect of precursor chemistry on the crystallisation and densification of sol–gel derived mullite gels and powders, *J. Sol–Gel Sci. Tech.* **10**, 19–30 (1997).

9. I.M. Low and R. McPherson, The Origins of Mullite Formation, *J. Mat. Sci.* **24** (3), 926–936 (1989).

10. A. Kumar Chakravorty and D. Kumar Ghosh, Synthesis and 980°C phase development of some mullite gels, *J. Am. Ceram. Soc.* **71** (11), 978–87 (1988).

11. J.P. Pollinger and G.L. Messing, Metastable solid solution extension of mullite by rapid solidification, *J. Mat. Res.* **3** (2), 375–379 (1988).

12. H. Schneider and T. Rymon-Lipinski, Occurrence of pseudotetragonal mullite, *J. Am. Ceram. Soc.* **71** (3), C162–C164 (1988).

13. M. Sales and J. Alarcon, Synthesis and phase transformations of mullites obtained from SiO_2–Al_2O_3 Gels *J. Eur. Ceram. Soc.* **16**, 781–789 (1996).

14. H. Schneider, R. X. Fischer, and D. Voll, Mullite with lattice constants $a > b$, *J. Am. Ceram. Soc.* **76** (7), 1879–1881 (1993).

15. R.X. Fischer, H. Schneider, and D. Voll, Formation of aluminum rich 9:1 mullite and its transformation to low alumina mullite upon heating, *J. Eur. Ceram. Soc.* **16**, 109–13 (1996).

16. N. L. Bowen and J. W. Grieg, "The System Al_2O_3–SiO_2," *J. Am. Ceram. Soc.* **7**, 238–54 (1924).

17. E.C. Shears and W.A. Archibald, Aluminosilicate refractories, *Iron Steel* **27**, 26–30 and 61–65 (1954).

18. N.A. Toropov and F. Ya. Galakhov, Solid solutions in the system Al_2O_3–SiO_2, *Izv. Aka. Nauk SSSR Otd. Khim. Nauk* **1958**, 8 (1958).

19. S. Aramaki and R. Roy, Revised equilibrium diagram for the system Al_2O_3, *J. Am. Ceram. Soc.* **45** (5), 229–222 (1962).

20. I.A. Aksay and J. A. Pask, Stable and metastable phase equilibria in the system Al_2O_3–SiO_2, *J. Am. Ceram. Soc.* **58** (11–12), 507–512 (1975).

21. T.J. Mroz, *Microstructural Evolution of Mullite Formed by Reaction Sintering of Sol-Gels*, MS Thesis, Dept. of Cer. Eng., Alfred University, Alfred, NY, 1988.

22. J.C. Huling and G.L. Messing, Epitactic nucleation of spinel in aluminum silicate gels and effect on mullite crystallization, *J. Am. Ceram. Soc.* **74** (10), 2374–2381 (1991).

23. S. Sundaresan and I.A. Aksay, Mullitization of diphasic aluminosilicate gels, *J. Am. Ceram. Soc.* **74** (10), 2388–2392 (1991).

24. M.D. Sacks, Y.-J. Lin, G.W. Scheiffele, K. Wang, and N. Bozkurt, Effect of seeding on phase development, densification behavior, and microstructure evolution in mullite fabricated from microcomposite particles, *J Am. Ceram. Soc.* **78** (11), 2897–2906 (1995).

25. R.F. Davis and J.A. Pask, Diffusion and reaction studies in the system Al_2O_3–SiO_2, *J. Am. Ceram. Soc.* **55** (10), 525–531 (1972).

26. S.H. Risbud and J.A. Pask, Mullite crystallization from SiO_2–Al_2O_3 melts, *J. Am. Ceram. Soc.* **61** (1–2), 63–67 (1978).

27. C.G. Bergeron and S.H. Risbud, *Introduction to Phase Equilibria in Ceramics*, Am. Ceram. Soc., Columbus, OH, 1984.

28. F.J. Klug, S. Prochazka, and R.H. Doremus, Alumina–silica phase diagram in the mullite region, *J. Am. Ceram. Soc.* **70**, 750–759 (1987).

29. J.A. Pask, The Al_2O_3–SiO_2 system: logical analysis of phenomenological experimental data, in *Ceramic Microstructures: Control at the Atomic Level*, A.P. Tomsia and A.M. Glaeser (eds.), Plenum, New York, NY, 1998, pp.255–262.

30. W.E. Worrall, *Clays and Ceramic Raw Materials*, Halsted Press, New York, NY, 1975, pp. 151–152.

31. D.G. Goski and W.F. Caley, Reaction sintering of kyanite and alumina to form mullite composites, *Can. Metall. Q.* **38** (2), 119–126 (1999).

32. S. Prochazka and F.J. Klug, Infrared-transparent mullite ceramics, *J. Am. Ceram. Soc.*, **66**, 874–880 (1983).

33. Y.M.M. Al-jarsha, H.G. Emblem, K. Jones, M.A. Mohd, A.B.D. Rahman, T.J. Davies, R. Wakefield, and G.K. Sargeant, Preparation, characterization and uses of mullite grain, *J. Mat. Sci.* **25** (6), 2873–2880 (1990).

34. R.D. Nixon, S. Chevacharoenkul, and R.F. Davis, Creep of hot-pressed SiC-Whisker-reinforced mullite, *Ceram. Trans.* **6**, 579–603 (1990).

35. H. Schneider, K. Okada, and J.A. Pask, *Mullite and Mullite Ceramics*, Wiley, New York, NY, 1994.

36. R.G. Chandran and K.C. Patil, A Rapid combustion process for the preparation of crystalline mullite powders, *Materi. Lett.* **10** (6), 291–295 (1990).
37. M. Schmuecker and H. Schneider, Structural development of single phase (type I) mullite gels, *J. Sol–Gel Sci. Tech.* **15**, 191–199 (1999).
38. M.J. Hyatt and N.P. Bansal, Phase transformations in xerogels of mullite composition, *J. Mat. Sci.* **25** (6), 2815–2821 (1990).
39. D.X. Li and W.J. Thomson, Mullite formation kinetics of a single-phase gel, *J. Am. Ceram. Soc.* **73** (4), 964–969 (1990).
40. P. Colomban and L. Mazerolles, SiO_2–Al_2O_3 phase diagram and mullite non-stoichiometry of sol–gel prepared monoliths: influence on mechanical properties, *J. Mat. Sci. Lett.* **9** (9), 1077–1079 (1990).
41. J.C. Huling and G.L. Messing, Hybrid gels for homoepitactic nucleation of mullite, *J. Am. Ceram. Soc.* **72** (9), 1725–1729 (1989).
42. K.J.D. MacKenzie, R.H. Meinhold, J.E. Patterson, H. Schneider, M. Schmuecker, and D. Voll, Structural evolution in gel-derived mullite precursors, *J. Eur. Ceram. Soc.* **16**, 1299–1308 (1996).
43. L.J. Andrews, G.H. Beall, and A. Lempicki, Luminescence of Cr^{3+} in mullite transparent glass ceramics, *J. Lumin.* **36** (2), 65–74 (1986).
44. P.A. Lessing, R.S. Gordon, and K.S. Mazdiyasni, Creep of polycrystalline mullite, *J. Am. Ceram. Soc.* **58** (3–4), 149–150 (1975).
45. W. Kollenberg and H. Schneider, Microhardness of mullite at temperatures to 1000°C, *J. Am. Ceram. Soc.* **72** (9), 1739–1740 (1989).
46. C. Paulmann, Study of oxygen vacancy ordering in mullite at high temperature, *Phase Trans.* **59**, 77–90 (1996).
47. T. Kumazawa, S. Ohta, H. Tabata, and S. Kanzaki, Influence of chemical composition on the mechanical properties of SiO_2–Al_2O_3 ceramics, *J. Ceram. Soc. Jpn.* **96**, 85–91 (1988).
48. Y. Okamoto, H. Fukudome, K. Hayashi, and T. Nishikawa, Creep deformation of polycrystalline mullite, *J. Eur. Ceram. Soc.* **6** (3), 161–168 (1990).
49. R. Torrecillas, J.M. Calderon, J.S. Moya, M.J. Reece, C.K.L. Davies, C. Olagnond, and G. Fantozzi, Suitability of mullite for high temperature applications, *J. Eur. Ceram. Soc.* **19**, 2519–2527 (1999).
50. H. Ohira, M.G.M.U. Ismail, Y. Yamamoto, T. Akiba, and S. Somiya, Mechanical properties of high purity mullite at elevated temperatures, *J. Eur. Ceram. Soc.* **16**, 225–229 (1996).
51. H. Schneider, "Thermal Expansion of Mullite," *J. Am. Ceram. Soc.* **73** [7] 2073–6 (1990).
52. I. Rommerskirchen, F. Cháveza, and D. Janke, Ionic conduction behaviour of mullite ($3Al_2O_3 \cdot 2SiO_2$) at 1400 to 1600°C, *Solid State Ionics* **74** (3–4), 179–187 (1994).
53. Y. Ikuma, E. Shimada, S. Sakano, M. Oishi, M. Yokoyama, and Z.E. Nakagawa, Oxygen self-diffusion in cylindral single-crystal mullite, *J. Electrochem. Soc.* **146** (12), 4672–4675 (1999).
54. P. Fielitz, G. Borchardt, M. Schmuecker, H. Schneider, and P. Willich, Measurement of oxygen grain boundary diffusion in mullite ceramics by SIMS depth profiling, *Appl. Surf. Sci.* **203–204**, 639–643 (2003).
55. Y.-M. Sung, Kinetics analysis of mullite formation reaction at high temperatures, *Acta Mater.* **48**, 2157–2162 (2000).
56. M.D. Sacks, K. Wang, G.W. Scheiffele, and N. Bozkurt, Effect of composition on mullitization behavior of α-alumina/silica microcomposite powders, *J. Am. Ceram. Soc.* **80** (3), 663–672 (1997).
57. B.R. Johnson, W.M. Kriven, and J. Schneider, Crystal structure development during devitrification of quenched mullite, *J. Eur. Ceram. Soc.* **21**, 2541–2562 (2001).
58. T. Huang, M.N. Rahaman, T.-I. Mah, and T.A. Parthasarathay, Anisotropic grain growth and microstructural evolution of dense mullite above 1550°C, *J. Am. Ceram. Soc.* **83** (1), 204–210 (2000).

Chapter 3
The Sillimanite Minerals: Andalusite, Kyanite, and Sillimanite

Richard C. Bradt

Abstract The chemistry and the mineralogy of the three $Al_2O_3 \cdot SiO_2$ sillimanite minerals (anadlusite, kyanite, and sillimanite) are described. Their P–T diagram is discussed. The structural differences among the three are reviewed, emphasizing the coordination of the Al^{3+} cations that link the double octahedral chains within the structures. Their decompositions to produce mullite and silica are described and contrasted. The effect of nanomilling on those decompositions is discussed. Finally, the locations of commercial deposits and the industrial applications are addressed.

1 Introduction

The sillimanite minerals are the three anhydrous aluminosilicates: andalusite, kyanite, and sillimanite [1,2]. Kyanite is also referred to as cyanita, cyanite, and disthene. Because all three have the same 1:1 molar ratio of alumina (Al_2O_3) to silica (SiO_2), they are often written simply as $Al_2O_3 \cdot SiO_2$ or Al_2SiO_5. Their ideal composition is 62.92 wt% alumina and 37.08 wt% silica. However, in natural states involving significant impurities, the alumina content is usually less than 60 wt%. There are reports of higher alumina content deposits associated with the presence of higher alumina content minerals. That such a mineral group exists is not surprising, for the three most common elements in the Earth's crust are O, Si, and Al.

The three sillimanite minerals are not found in phase diagram of the familiar binary alumina–silica at a pressure of 1 atm. This phase diagram has only the single aluminosilicate compound known as mullite, $3Al_2O_3 \cdot 2SiO_2$ (71.79 wt% alumina and 28.21 wt% silica) [3–5]. The absence of the three sillimanite minerals on the binary diagram is because they are geologically formed at high pressures and elevated temperatures within the earth. None of the three minerals of the sillimanite group are equilibrium phases at 1 atm pressure. However, the three do exist in their metastable states throughout the world, quite abundantly in many geologically favorable areas. They are often discussed simultaneously with mullite, for their chemistries and crystalline structures are related to that of mullite. Furthermore, the three sillimanite polymorphs form mullite by decomposition when heated to elevated temperatures. This characteristic is the basis of their industrial utility.

J.F. Shackelford and R.H. Doremus (eds.), *Ceramic and Glass Materials:*
Structure, Properties and Processing.
© Springer 2008

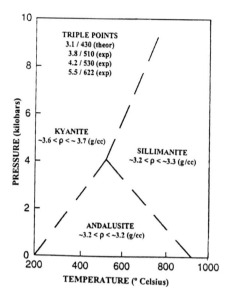

Fig. 1 The *P–T* diagram of Al$_2$O$_3$·SiO$_2$ in the vicinity of the triple point

The *P–T* diagram of the three sillimanite minerals is presented in Fig. 1. This *P–T* diagram for the Al$_2$SiO$_5$ compound is somewhat controversial to the precise *P–T* values for the equilibrium triple point of the three polymorphs. Four sets of triple point *P–T* data, three experimental and one theoretical, are listed in this diagram. It is evident that the triple point is located at approximately 3–6 kbar pressure and a temperature of approximately 400–650°C. Kyanite is the high-pressure structure of the three, while sillimanite is the high-temperature structure. Andalusite occurs at lower pressures and temperatures. When the three minerals are heated above approximately 1,200°C in air, they will decompose to produce the stable compound mullite and reject very fine, highly reactive silica. The decomposition temperatures vary for the three sillimanite minerals as does the structural form of the rejected silica. The decompositions will be discussed further below.

It is important to address a few issues regarding the three polymorph densities as they appear in the *P–T* diagram of Fig. 1. Generally, high temperature polymorphs of materials are less dense than low temperature ones. However, this is not the case when andalusite (~3.1 < ρ < ~3.2 Mg m^{-3}) is compared with sillimanite (~3.2 < ρ < ~3.3 Mg m^{-3}). In this comparison, the high temperature form, sillimanite, is denser. This apparent anomaly occurs because of a change in the coordination number of a portion of the Al cations within the crystal structure. The coordination change is from the unusual value of fivefold within andalusite to just fourfold within sillimanite. The density effects of pressure are to be expected with the high-pressure polymorph, kyanite, having the highest density (~3.5 < ρ < ~3.7 Mg m^{-3}).

A number of properties [6] of the sillimanite group minerals and those of mullite as well are summarized in Table 1. These properties merit brief discussion. The three minerals have relatively large unit cells and complex crystal structures. Their crystal habits reflect their easy and well-defined cleavage planes. They are not very hard minerals by any measure, typically in the middle of the Mohs hardness scale. Perhaps

Table 1 Some characteristics of the sillimanite minerals and mullite

Properties	Kyanite	Andalusite	Sillimanite	Mullite
Formula	$Al^{VI}Al^{VI}Si^{IV}O_5$	$Al^{VI}Al^{V}Si^{IV}O_5$	$Al^{VI}Al^{IV}Si^{IV}O_5$	$Al^{VI}(Al_{1+2x}Si_{1-2x})^{IV}O_5$
Crystallography	Triclinic	Orthorhombic	Orthorhombic	Orthorhombic
Unit cell (nm)				
a	0.712[a]	0.779	0.748	0.754
b	0.785	0.790	0.767	0.769
c	0.557	0.555	0.577	0.289
Crystal habit	Lathlike	Prismatic	Fibrous	Acicular
Density (Mg m⁻³)	3.5~3.7	3.1~3.2	3.2~3.3	3.1~3.3
Mohs Hardness	5½~7	6½~7½	6½~7½	6~7
Cleavage	{100} pf {010} gd	{110} gd	{010} pf	{010} pf
Refractive index				
α	1.710–1.718	1.633–1.642	1.653–1.661	1.630–1.670
β	1.719–1.724	1.639–1.644	1.657–1.662	1.636–1.675
γ	1.724–1.734	1.644–1.650	1.672–1.682	1.640–1.690
$T_{decomposition}$	~1,150–1,350°C	~1,250–1,500°C	~1,400–1,700°C	N/A
$T_{decomp.interval}$	~200°C	~250°C	~300°C	N/A
$\Delta V_{decomposition}$	~+15%	~+4%	~+8%	N/A
Rejected SiO_2	Cristobalite	Amorphous	Amorphous	N/A

[a]The triaxial angles for the triclinic Kyanite are $\alpha = 89.98°$, $\beta = 101.12°$, and $\gamma = 106.01°$

their most distinguishing feature is the systematic change of coordination of the Al^{3+} cations from one structure to another. Only kyanite, which is triclinic, has a unit cell in which the axes are not orthogonal.

2 Structures of the Sillimanite Minerals

The three sillimanite minerals are structurally similar and have structures that are related to that of mullite. It is not surprising that they all form mullite upon decomposition. Kyanite crystallizes in the triclinic system, while sillimanite, andalusite, as well as mullite have orthorhombic crystal structures. In these structures, all the Si^{4+} cations are in fourfold coordination with O^{2-} anions, but the Al^{3+} cations exist in four-, five-, and sixfold coordination with O^{2-} anions, and therein lie the structural differences. The fivefold coordination of some Al^{3+} cations within AlO_5 polyhedra is rather unusual, perhaps the result of formation at high pressures. The other structural differences among the three minerals are quite small. They are associated with the double chain structures of these three minerals and the linkages of the chains to one another by different alumina and silica polyhedra. Those concepts are readily extended to mullite.

In sillimanite itself, where the Al^{3+} cations are in four- and sixfold coordination, double chains of aluminum oxide octahedra exist parallel to the c-axis of the orthorhombic structure. These chains are formed by the edge sharing of the AlO_6 octahedra. Those chains are linked or connected by alternating AlO_4 and SiO_4 tetrahedra. Note the uniqueness of the presence of Al^{3+} cations in fourfold coordination, which is again a consequence of the mineral formation at high pressures. The coordination of the Al^{3+} cations in sillimanite is evenly divided. Half of the Al^{3+} cations are in tetrahedral and half are in octahedral coordination. As one considers andalusite and then kyanite, it is the coordination of the Al^{3+} cations that characterizes the structural differences. In andalusite, which is also orthorhombic in structure, the AlO_6 octahedra chains are linked by SiO_4 tetrahedra and AlO_5 polyhedra alternate within the structure. Proceeding to kyanite, the chains are linked by SiO_4 tetrahedra and AlO_6 octahedra. It is evident that the major structural difference for the three sillimanite minerals is the Al^{3+} cation coordination that forms the linkages between the double AlO_6 octahedra chains. In all three minerals, the Si^{4+} cations are always tetrahedrally coordinated with oxygens, while half of the Al^{3+} cations are always octahedrally coordinated with oxygens. It is the coordination of the other half of the Al^{3+} cations that changes in these structures. In sillimanite, the other half of the Al^{3+} cations are in fourfold or tetrahedral coordination, but, in anadalusite, they are in fivefold coordination. In kyanite, they are in sixfold or octahedral coordination. This critical change of the Al^{3+} cation coordination in the three minerals occurs in those polyhedra that crosslink the double chains that are formed by the edge sharing of the AlO_6 octahedra. Such distinctive structural differences can be considered to be a manifestation of the high pressures involved. The cation coordination numbers are presented as the superscript Roman numerals on the mineral chemical formulae in Table 1.

The structure of mullite is similar to that of the sillimanites, consistent with the fact that they decompose to form mullite at high temperatures and 1 atm pressure. It has been suggested that the double AlO_6 octahedral chain structure is preserved during the decomposition. The mullite structure is, however, somewhat complicated by its extensive stability over a wide range of stoichiometries. The composition of mullite can be expressed as

$$Al\left(Al_{1+2x}Si_{1-2x}\right)O_{5-x} \qquad (1)$$

where the value of x may vary from $\sim 0.08 < x < \sim 0.29$. It is evident that the common 3:2 mullite is not the only stoichiometry. The 2:1 mullite structure is common in fusion cast mullites. In mullite, the AlO_6 octahedra also form double chains parallel to the c-axis of the orthorhombic structure and are also cross-linked by alumina and silica tetrahedra. This is one reason why mullite is often considered along with the sillimanite minerals in the literature. These structural aspects of mullite, along with those of the three sillimanite minerals, are summarized in Table 1, which allows for their direct comparisons on the basis of several different physical properties.

3   Decomposition of the Sillimanite Minerals

As the sillimanite minerals are geologically formed at high pressures, they decompose or undergo a structural decomposition when heated to elevated temperatures in air at 1 atm pressure [7–10]. The decomposition reaction can be written as

$$3\,Al_2O_3 \cdot SiO_2 = 3\,Al_2SiO_5 \rightarrow 3Al_2O_3 \cdot 2SiO_2 + SiO_2 \qquad (2)$$

where the 3:2 stoichiometric mullite and silica are the decomposition products. The previously mentioned 2:1 mullite does not form during the decomposition of the sillimanites. The form of the silica varies for the three polymorphs. For kyanite, the decomposition product is cristobalite, while for andalusite and sillimanite, it is amorphous. It is worth noting that the temperature ranges of the decompositions are compatible with cristobalite formation. The decompositions of kyanite and andalusite are topotactical, but for sillimanite, the process involves multiple steps with the intermediate formation of a disordered sillimanite structure. Numerous similarities and differences exist for the decomposition reactions.

Extensive attrition milling of the original minerals to the nanosize range alters the form of the resulting silica to cristobalite. In all instances, the resulting silica is highly reactive. Because of this state of the silica, addition of reactive aluminas to the sillimanite minerals before heating to the decomposition temperature results in the immediate reaction with the rejected silica to form a secondary mullite in addition to the primary mullite from the original sillimanite mineral. Complete mullitization of a pure sillimanite mineral requires the addition of 31.42 wt% alumina. It is possible to produce pure single-phase mullite bodies through this technical approach. When fired properly, they can be sintered to a dense, fine grain size ceramic body.

The decompositions of the three sillimanite minerals occur over different temperature ranges and with different volumetric expansions. As the sillimanites are formed at high pressures, it is natural that they exhibit large volumetric expansions when they decompose at 1 atm pressure. As expected, kyanite, the highest-pressure form, undergoes the largest volumetric expansion (about 15%). The P–T formation densities can be viewed as the driving force for the decompositions. Kyanite also initiates its decomposition at the lowest temperature of the three.

As expected for any kinetic process, the sillimanite decompositions are both temperature and time dependent. Typically, when slowly heated, kyanite begins to decompose at ~1,150°C and has completely decomposed by ~1,350°C. Andalusite starts its decomposition at ~1,250°C and is fully transformed by ~1,500°C. Sillimanite is less prone to decomposition and first begins to decompose at ~1,400°C and is not fully decomposed until ~1,700°C. The temperature intervals over which the decompositions occur increase in the following order: kyanite, andalusite, sillimanite. They are approximately 200°C, 250°C, and 300°C, respectively. The decomposition temperatures and intervals are specified as "about" or "approximate" in every instance. The three minerals will themselves vary slightly depending on the specific geological origins that determine their exact location within the original P–T fields in Fig. 1. For that reason, the densities and the driving force for their decompositions vary, even for the same mineral from different locations.

The volume expansions of the decompositions can be beneficial and at the same time a hindrance to their industrial applications. Kyanite has the largest $+\Delta V\%$ and is often as large as 15% or even slightly greater. Firing kyanite by itself will result in the individual crystal prisms of the mineral bloating and cracking severely. Of course, this macrostructural destruction of the crystals is highly beneficial to any subsequent milling and homogenization in technical ceramic bodies and refractories. It does, however, somewhat restrict the utilization of "pure" kyanite, as mined, for a high temperature ceramic body. The decomposition volume expansions of andalusite and sillimanite are both somewhat less than that of kyanite, ascribing to the relative pressure levels of their equilibrium formation. The volume expansion of andalusite is only about 4%, while that of sillimanite is larger at about 8%.

That the three sillimanites should exhibit large volume expansions upon decomposition is not surprising when the densities of the products of the decomposition are considered. Mullite, the major product, has a density of about $3.2\,Mg\ m^{-3}$, similar to that of andalusite. Thus, andalusite experiences the lowest of the volume expansions during its decomposition. The crystalline silica phase that results from the kyanite decomposition is cristobalite with a density of 2.2–$2.3\,Mg$ m^{-3} and the amorphous silica would be expected to be even less; thus, it is easy to explain the volume expansion and cracking of the sillimanite minerals when they decompose, purely on the basis of the densities of the decomposition products compared with that of the original sillimanite mineral.

When the sillimanite minerals are milled into the nanoparticle regime, the decompositions are strongly modified in several ways. The first effect is that the temperature range of the decomposition to mullite and silica is reduced significantly, often by several hundred degrees Celsius. This decrease in temperature has potential energy savings for subsequent industrial firing processes. Second, the volume expansion, $\Delta V\%$, is reduced, eventually achieving a situation where the sintering of the nanofine particles supercedes the volume expansion of the decomposition and shrinkage occurs upon heating. Finally, the products of the decomposition are altered, as the amorphous silica of the andalusite and sillimanite decompositions becomes cristobalite comparable to the decomposition of kyanite. Although there have not been high-resolution transmission electron microscopy studies of the effects of fine milling on the decomposition, it is highly probable that the detailed mechanisms of the decomposition of very fine sillimanite mineral particles may be altered from those of coarse particles usually produced industrially.

4 Worldwide Deposits of the Sillimanite Minerals

Because of the abundance of O, Si, and Al in the earth's crust, it is not surprising that there exist numerous deposits of the sillimanite minerals worldwide. However, the qualities, or grades, of most of these deposits are not amenable to large-scale commercial mining and beneficiation. Many are just small quarrying operations. For those reasons, only a few of the major worldwide deposits will be noted. The text by Varley [2] has an excellent summation of the deposits worldwide, while Chang [1] addresses the mineral processing and beneficiation quite well.

The major industrial sources of andalusite are in South Africa. The Transvaal andalusite often approaches 60 wt% alumina and so the quality is very high indeed. Sometimes, these ores contain other high alumina minerals, so that the total alumina content may actually exceed the 62.9 wt% level of pure anadalusite. Many other deposits exist, notably in the United States and the former USSR, but they are not of the significance of the South African ones. Of course, new findings are always possible in other localities.

The major kyanite deposits are in the United States and India. The US deposits are in the eastern states along the Atlantic seaboard. They are of exceptionally high quality. Massive kyanite deposits also occur in India, where the mining is extensive as they too are excellent. Other less productive deposits occur in Africa, Australia, South America, and the former USSR.

Sillimanite occurs to some extent almost everywhere that andalusite and kyanite are found. However, the major commercial sources are in India and in the sillimanite beach sands of several locales. The Assam deposits in India are perhaps the most famous, as large pure sillimanite boulders are found there. Beach sands in Australia, the United States, and Africa are also sources, often as a byproduct of the beneficiation of the heavy metals in those sands.

Although the sillimanite minerals are the major source of mullite for ceramics and refractories, there was an increase in the production of synthetic mullites by electrofusion starting in the 1950s and shortly thereafter by sol–gel chemical methods. These industries were the outgrowth of the inability of the commercial sillimanite producers to provide sufficiently high quality amounts of the sillimanite minerals to meet the worldwide demand for mullite. Synthetic mullite was the result. However, the use of synthetic mullite in many applications requires extensive manufacturing process alterations when substituted for the sillimanites. For example, fused mullite is often of the $2Al_2O_3 \cdot SiO_2$ variety, although it is possible to produce different stoichiometries with different alumina contents. Chemical mullites prepared by sol–gel methods are highly pure, but they are quite expensive relative to the varieties naturally produced from the sillimanites.

5 Industrial Applications of the Sillimanite Minerals

It has already been noted that the primary use of the sillimanite minerals is for the production of mullite refractories. These refractories are used extensively in the steel and glass industries. Because of the association of the sillimanite minerals with higher alumina content minerals in their deposits, the range of alumina contents for the aluminosilicate refractories that they produce is approximately 50–80 wt% alumina.

There has also been a modest use of the sillimanite minerals as abrasives. This was originally a matter of convenience, perhaps even mistaking them for corundum. However, the recent globalization of trade has practically eliminated this application, given the modest hardness of the sillimanites, only about 7 on the mineralogical Mohs scratch hardness scale. This hardness level cannot compete with corundum and the recently developed and quite superior synthetic abrasives.

As with many other minerals, the sillimanites are sometimes found in large single crystal form at the quality level suitable for gemstones. This is particularly true for andalusite. Exquisite green andalusite is common in Brasil in the state of Minas Gerais. It can frequently be obtained at gem and mineral shows around the world.

6 Conclusions

The structures and properties of the three sillimanite minerals (andalusite, kyanite, and sillimanite) have been discussed. The uniqueness of the three, perhaps the result of their high-temperature, high-pressure genesis is described. Their decompositions at 1 atm pressure are also described. The availability and the properties of these minerals have served them well in industrial applications in the past. It appears that their worldwide source locations provide them with numerous opportunities for future uses in a growing market.

References

1. L.L.Y. Chang, Sillimanite, Andalusite, and Kyanite, in *Industrial Mineralogy*, Prentice-Hall, Upper Saddle River, NJ, 2002, pp. 373–385.
2. E.R. Varley, *Sillimanite: Andalusite, Kyanite, Sillimanite*, Chemical Publishing, New York, NY, 1968.
3. J. Grofcsik, *Mullite, Its Structure, Formation, and Significance*, Hungarian Academy of Sciences, Budapest, 1961.
4. H. Schneider, K. Okada, and J.A. Pask, *Mullite and Mullite Ceramics*, John Wiley, Chichester, England, 1994.
5. M.J. Hibbard, *Mineralogy*, McGraw-Hill, New York, NY, 2002, pp. 49, 425, 459, and 479.
6. J.K. Winter and S. Ghose, Thermal expansion and high-temperature crystal chemistry of the Al_2SiO_5 polymorphs, *Am. Mineral.* **64**, 573–586 (1979).
7. J. Aguilar-Santillan, R. Cuenca-Alvarez, H. Balmori-Ramirez, and R.C. Bradt, Mechanical activation of the decomposition and sintering of kyanite, *J. Am. Ceram. Soc.* **85**, 2425–2431 (2002).
8. H. Schneider and A. Majdic, Kinetics and mechanism of the solid state high temperature transformation of andalusite (Al_2SiO_5) into 3/2 mullite, $3Al_2O_3 \cdot 2SiO_2$, and silica, SiO_2, *Ceramurgia Int.* **5**, 31–36 (1979).
9. H. Schneider and A. Majdic, Kinetics of the thermal decomposition of kyanite, *Ceramurgia Int.* **6**, 32–37 (1980).
10. H. Schneider and A. Majdic, Preliminary investigation on the kinetics of the high-temperature transformation of sillimanite to 3/2 mullite, $3Al_2O_3 \cdot 2SiO_2$, and silica, SiO_2, and comparison with the behavior of andalusite and kyanite, *Sci. Ceramics* **11**, 191–196 (1981).

Chapter 4
Aluminates

Martin C. Wilding

Abstract Aluminates form in binary systems with alkali, alkaline earth or rare-earth oxides and share the high melting point and resistance to chemical attack of the pure Al_2O_3 end-member. This means that these ceramics have a variety of applications as cements, castable ceramics, bioceramics, and electroceramics. Calcium aluminate cements are used for example in specialist applications as diverse as lining sewers and as dental restoratives.

Ceramics in aluminate systems are usually formed from cubic crystal systems and this includes spinel and garnet. Rare earth aluminate garnets include the phase YAG (yttrium aluminium garnet), which is an important laser host when doped with Nd(III) and more recently Yb(III). Associated applications include applications as scintillators and phosphors.

Aluminate glasses are transparent in the infrared region and these too have specialist applications, although the glass-forming ability is poor. Recently, rare earth aluminate glasses have been developed commercially in optical applications as alternatives to sapphire for use in, for example, infrared windows.

Aluminates are refractory materials and their synthesis often simply involves solid-state growth of mixtures of purified oxides. Alternative synthesis routes are also used in specialist applications, for example in production of materials with controlled porosity and these invariably involve sol–gel methods. For glasses, one notable, commercially important method of production is container-less synthesis, which is necessary because of the non-Arrhenius (fragile) viscosity of aluminate liquids.

1 Introduction

Glasses and ceramics based on the Al_2O_3-based systems have important applications as ceramic materials, optical materials, and biomedical materials. Aluminate materials include alkaline earth aluminates, such as those in the $CaO–Al_2O_3$ system, which are refractory cousins of hydrous Portland cement [1–3]. Calcium aluminates have a role as both traditional ceramic and cement materials and are used for example as refractory cements; however, calcium aluminates are also important for more novel applications

J.F. Shackelford and R.H. Doremus (eds.), *Ceramic and Glass Materials:*
Structure, Properties and Processing.
© Springer 2008

such as bioceramics [4–8]. Rare earth (yttrium and the lanthanides) aluminates are important laser host materials. Yttrium aluminum garnet (YAG) is one of the most common laser hosts; Nd-doped YAG lasers with powers of up to 5 kW are important for welding and cutting applications and have the further advantage of being solid state, the primary laser component being a single crystal of Nd-doped YAG. Associated with the laser properties of YAG are the materials characteristics of rare earth aluminates, which favor applications as refractory ceramics, composite laser hosts, and glass fibers that are important for optical applications, but also can be used in composite materials [9, 10].

Many of the important desirable properties that make aluminates important in materials science are similar to those of the end-member Al_2O_3. This includes the refractory nature of aluminates, for example Al_2O_3 melts at 2,054°C and other important aluminates have similarly high melting points (Table 1). In addition, aluminates have high hardness, high strength, and are resistant to chemical attack. Al_2O_3 and both calcium and rare earth aluminate systems can have useful properties such as transparency in the infrared region, and this makes aluminate glasses important for use as optical fibers. Because of their optical applications, aluminate glasses have been studied extensively and as a consequence some very unusual and anomalous thermodynamic properties have come to light.

The refractory nature of aluminates means that high temperature synthesis techniques are required. Depending on the application, aluminates can be made by mixing of oxides and subjecting the mixtures to high temperature, as for example in the manufacture of cement. For other applications, such as optical uses, more exotic techniques are used. These include high temperature melting, single crystal growth [11, 12], container-less synthesis of glasses using levitation [13], and low-temperature routes such as sol–gel synthesis [14, 15] and calcining.

There are a variety of important crystal structures in aluminate systems. Among the most important are the spinel [16] and garnet structures [17, 18]. These various structures reflect differences in the coordination polyhedron of both Al(III) and added components such as Mg(II), Ca(II), and the rare earth ions. In addition, studies of glass structure suggest a wealth of different coordination environments for both Al(III) and added components and structures that are not simply disordered forms of crystalline phases.

For the purposes of this review, aluminates can be defined as a binary section of a ternary oxide system with Al_2O_3 as one component. A large number of different aluminates can be made and it is not the purpose of this chapter to provide an exhaustive list of each different aluminate type or each application. Rather, it is the purpose of this chapter to provide a survey of the range of binary Al_2O_3-systems and to demonstrate the diversity of both their applications to materials science and to elaborate on the unusual

Table 1 The physical properties of selected binary aluminate ceramics

	Melting temperature (K)	Density (g cm^{-3})	Hardness (Knoop/100 g) (Kg mm^{-2})	Compressive strength (MPa)	Tensile strength (MPa)	Young's modulus (GPa)
α-Al_2O_3	2,327	3.98	2,000–2,050	2,549	255	393
$CaAl_2O_4$	2,143	2.98				
$MgAl_2O_4$	2,408	3.65	1,175–1,380	1,611	129	271
$LiAlO_2$	1,883	2.55			350	
$Y_3Al_5O_{12}$	2,243	4.55	1,315–1,385		280	282

structural and thermodynamic properties of crystalline and glassy aluminate materials. Three subsets of aluminates will be highlighted: binary alkaline earth aluminates (CaO, MgO–Al_2O_3), which includes calcium aluminate cements and magnesium aluminate spinels, alkali (lithium) aluminates, which potentially have very important applications in the development of new types of nuclear reactors [19], and rare earth aluminates, particularly compositions close to that of the yttrium aluminum garnet ($Y_3Al_5O_{12}$; YAG). Each section will discuss the applications, structure, and synthesis of each composition, and finally the thermodynamic and structural properties of these aluminates will be compared and summarized.

2 Alkali and Alkaline Earth Aluminates

There are two important systems discussed in this section: CaO–Al_2O_3 and MgO–Al_2O_3. In addition, there are sometimes other binary systems and mixtures or small amounts of additional elements added to the binary systems such as SrO, added to improve glass forming ability [20, 21]. For convenience, these latter more complex systems will be discussed with the strict binary CaO–Al_2O_3 phases.

3 Calcium Aluminate Cements

Calcium aluminate phases are used as cements in refractory and other specialized applications [1, 2, 22, 23]. The ceramics in the calcium aluminate (CaO–Al_2O_3) system are closely related to Portland cements and have similar properties in terms of rapid hardening and setting times [24]. Their phase equilibria are closely related to that of Portland cements as and are formed in the binary CaO–Al_2O_3 of the ternary CaO–Al_2O_3–SiO_2 phase diagram. The binary phase diagram (Fig. 1) shows that calcium aluminate cements (CACs) have a wider range of compositions than Portland cements, but are dominated by the monocalcium aluminate ($CaAl_2O_4$, also referred to as CA).

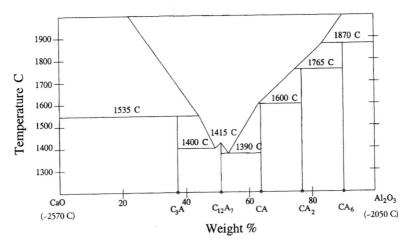

Fig. 1 The CaO–Al_2O_3 phase diagram [25, 26]

CACs were developed in response to the need for cements resistant to groundwater and seawater attack and are the only cements, other than Portland cement, that are in continuous long-term production [2]. The property of CAC that was most important in their commercial development is the resistance to sulfate attack, which contrasted with the poor-sulfate resistance of contemporary Portland cements [2], and CAC was first patented in 1908 [2]. Most early applications, in construction projects following the First World War, were in structures exposed to seawater, such as harbor pilings. Because CAC hardens rapidly, it was adopted for prestressed concrete beams in the post World War II construction boom, with some unfortunate results. Poor understanding of the material properties of CAC and incorrect water to cement ratios led to the collapse of several buildings, and the use of Portland cements, which are cheaper, has replaced CAC in prestressed concrete beams[2].

There are, however, several important niche applications for CAC. Most notably, CACs are used as linings to sewers and mine tunnels. Calcium aluminate cements are resistant to chemical attack from sulfate-producing bacteria that thrive in sewer systems (especially in warmer climates), and sprayed concrete linings to sewers have been shown to resist degradation for periods up to 30 years. The high impact and abrasion resistance of CAC also makes it suitable as a lining material for ore tunnels in mines and because CAC sets rapidly, it can be sprayed onto tunnel walls (as "shotcrete") and even used as a tunnel lining.

Additional specialist applications include castable refractory ceramics and use as bioceramics, which are discussed later.

4   Phase Equilibria and Crystal Phases in the $CaO–Al_2O_3$ System

The binary phase diagram of $CaO–Al_2O_3$ shows two refractory end-members CaO and Al_2O_3 with melting points of 2,570°C and 2,050°C, respectively [25, 26]. There is a deep eutectic with a minimum at 1,390°C and five intermediate crystalline phases, of which three hydrates are important as cements [27].

Monocalcium aluminate ($CaAl_2O_4$) is the most important phase in CAC. Addition of water to $CaAl_2O_4$ (CA) eventually leads to the formation of the crystalline hydrates $3CaO·Al_2O_3·6H_2O$ and $Al_2O_3·3H_2O$, which dominate the initial hydration of CAC [27]. Monocalcium aluminate $CaAl_2O_4$ does not have a spinel structure, even though it is stochiometrically equivalent to Mg-aluminate spinel. The crystal structure of this phase is monoclinic, pseudo hexagonal with a p2/n space group. The structure of the CA phase resembles that of tridymite and is formed from a framework of corner-linked AlO_4 tetrahedra. Large Ca^{2+} ions distort the aluminate framework, reducing symmetry. As a consequence, the coordination environment of Ca^{2+} is irregular.

The CA2 phase ($CaO·2Al_2O_3$) occurs as the natural mineral grossite [27]. This phase is a monoclinic C2/2 phase and is also formed from a framework of corner-linked AlO_4 tetrahedra. Some of the oxygens in the framework are shared between two tetrahedral and some are shared between three. The CA2 phase does not react well with H_2O and is not necessarily useful in refractory CAC. The CA6 phase also occurs naturally, as the mineral hibenite. This phase has a similar structure to β-Al_2O_3 and is nonreactive and its presence is not desired in CAC.

The C12A7 phase reacts very rapidly with water and becomes modified to produce the hydrated phase $11CaO \cdot 7Al_2O_3 \cdot Ca(OH)_2$. The C12A7 phase is cubic with a space group of 143 d. The basic structure is one of a corner shared AlO_4 framework. The Ca^{2+} ions are coordinated by six oxygen atoms but the coordination polyhedron is irregular. It has been suggested, by infrared spectroscopy, that some of the aluminum ions are coordinated by five oxygen atoms. This hydrated phase $(Ca_{11}Al_7 \cdot Ca(OH)_2)$ is closely related to the naturally occurring mineral, Mayenite, a cubic mineral with M2M symmetry and a large (11.97 Å) unit cell, closely related to the garnet structure.

5 Refractory Castables

One very important niche application for calcium aluminate (cements) is as refractory castables. Key to the success of calcium aluminates in this application are their refractory properties that contrast with those of Portland cements. Although Portland cement maintains good strength when heated, reactive components (CaO) are liberated and can absorb moisture from the atmosphere when cooled, causing expansion and deterioration of, for example, kiln linings. CACs are not much susceptible and can be used to form monolithic castables and refractory cements [28, 29].

6 Calcium Aluminate Bioceramics

Ceramic materials with high strength, high wear resistance, and high resistance to corrosion can be used as prosthetic replacements for bones and teeth. One important consideration for potential bioceramics is compatibility with the human body, since for example hip prostheses are placed in vivo. Bones and teeth comprise hydroxyapatite, a calcium bearing phase and Ca^{2+} ions are mobile during formation. Calcium aluminates are attractive for bioceramic applications; because of the mobility of Ca^{2+} in biological fluids these cements can bond to bone and are quick setting and during hardening form enough hydrates to fill the initial porosity and result in a high strength end-product. In addition, for dental applications CAC have similar thermal properties to teeth and are translucent, therefore even on the basis of aesthetic appearance are useful as a dental restorative [5–8, 30, 31].

7 Synthesis of Calcium Aluminates

CAC require large industrial facilities, similar to those used to make ordinary Portland cement. The raw materials for CAC are typically bauxite and limestone, which are ball-milled and mixed together to form a feed of appropriate composition, which is fed into rotary kilns to form a calcium aluminate clinker. The clinker is ball-milled to produce the cement. Analysis for composition and mineralogy at various stages of manufacture are essential to ensure a consistent product, see for example Chakraborty and Chattopadhyay [32] for a discussion of the bulk processing of high alumina CAC.

For high purity calcium aluminate compositions, solid-state synthesis is still the norm [33, 34]. Most CAC compounds are made by solid-state reactions between ground powders of calcium carbonate and purified alumina. The sintering temperatures depend on alumina content. More recently attempts have been made to synthesize CA compounds using processes with temperatures less than 900°C. These latter methods include sol–gel synthesis and precipitation and are important for production of high-purity homogenous powders with small grain size.

Amorphous calcium aluminate powders have been synthesized chemically by Uberoi and Risbud [35] by sol–gel methods. These materials were made from calcium nitrate ($Ca(NO_3)_2$) and by using aluminum di-sec-butoxide acetoacetic ester chelate ($Al(OC_4H_9)_2(C_6H_9O_3)$) as the source of alumina.

A further synthesis method is self-propagating combustion synthesis [33, 36, 37]. In this alternative approach, nitrate starting powders are dissolved in H_2O and urea (CH_4N_2O) is added. When this mixture is boiled, dehydrated, and dried, it forms a hygroscopic precursor to calcium aluminates, which can be crystallized by heating in dry air between 250 and 1,050°C. The gaseous decomposition products of the precursor mixture are NH_4 and HCNO, which ignite at ~500°C, locally the temperature in the dried foam increases to ~1,300°C, which promotes crystallization of the CAC phase.

8   Calcium Aluminate Glasses

Calcium aluminate glasses have the potential for a variety of mechanical and optical applications; [20, 21, 38–46] however, they are difficult to form. Addition of SiO_2 can be used to improve glass-forming ability, although this reduces the optical properties, particularly the transparency to infrared, so it is best avoided. Studies show that the best glass-forming composition in the $CaO–Al_2O_3$ binary is close to the composition $64CaO–36Al_2O_3$ [45].

Calcium aluminate glasses form from "fragile" liquids [47], and these deviate from an Arrhenius viscosity–temperature relation. Because of these distinct rheological properties, calcium aluminate glasses have been extensively studied by diffraction and spectroscopic techniques. The composition-dependence of calcium aluminate structures was studied by McMillan for almost the entire range of $CaO–Al_2O_3$ liquids [45] using extremely rapid quench techniques. Extensive NMR and Raman data obtained from these rapidly quenched glasses show a range in Al–O coordination. For $CaO/Al_2O_3 < 1$, the glasses are dominated by [IV]Al. NMR and Raman data indicate that there are changes in mid-range order and also in relaxation time (i.e., viscosity), as expected for fragile liquids [45]. The changes in Raman and NMR spectra are interpreted as different degrees of distortion of the Al–O coordination polyhedron as the identity of next-nearest neighbor changes. Raman data support this interpretation, in that there is no evidence for change in AlO_4 polymerization. Similarly, X-ray absorption spectroscopy shows dramatic changes in spectra with quench rate, and changes in next-nearest neighbor. For calcium aluminates it is argued that the rearrangement of next-nearest neighbors reflects over- and under-bonding of the central ion in the Al–O coordination polyhedron, dependent on the degree of distortion.

Neutron and combined neutron and X-ray diffraction data for 64:36 and 50:50 calcium aluminate glasses [40, 48] have been used to determine Al–O and Ca–O coordination environments and mid-range order changes. These studies show that the Al–O correlation at 0.176 nm and the area below this peak yield a first-neighbor coordination number of 4.8. There is a peak in the pair-correlation function at 0.234 nm, which corresponds to Ca–O; the area beneath this peak yields a coordination number of 4.0, inconstant with the value obtained from the radial distance by bond–valence theory [48]. Further examination of the diffraction data reveals a second Ca–O distance at 0.245 nm. Combined diffraction data suggest that the Ca–O polyhedron is quite distorted [40] and that the glass consists of a corner-shared Al–O framework with the Al–O units corner- and edge-shared with distorted Ca–O polyhedra.

9 Synthesis of Calcium Aluminate Glasses

Calcium aluminate glasses can be made using a variety of techniques, depending on the composition required. The ease of devitification is a considerable concern if calcium aluminate glasses are to be used for optical purposes. Although a strong network former such as SiO_2 can be added to improve glass-forming ability, this has detrimental effects on the optical properties.

Calcium aluminate glasses can be made quite easily by air quenching liquids of 61:39 composition, which is the composition most extensively used for glass fiber production [20, 44]. The glass-forming ability is considerably enhanced by adding components such as BaO, SrO, and NaO [21] without affecting the optical performance.

A further method of glass synthesis is container-less levitation techniques [39, 40, 49]. In this method, a ceramic precursor of appropriate composition is levitated by a gas jet and laser heated. Samples up to 4-mm diameter can be levitated in this way and because there is no container, heterogeneous nucleation is avoided. This means that liquids can be supercooled considerably and glasses formed from compositions are generally considered to be poor glass-formers, this includes calcium aluminate glasses. Fibers can be extracted from the levitated bead by using a tungsten "stinger" [13].

10 MgO–Al$_2$O$_3$ Aluminates

Magnesium aluminate phases have high melting points and like calcium aluminates are used in refractory ceramic applications. These applications include the linings of ladles in steel plants and linings for cement kilns. In these applications, ceramics are used either in the form of castables, in case of linings to labels, or as bricks (kiln linings) [50–57]. Having low phonon energy and good mechanical properties, magnesium aluminates are also emerging as an infrared window material [20].

The only stable compound in the MgO–Al$_2$O$_3$ system [58, 59] is spinel [16, 60] (MgAl$_2$O$_4$), which has a melting point of 2,105°C and in addition to being a refractory compound has high resistance to chemical attack and radiation damage [56, 61–64]. Spinel ceramics have potential use for a variety of applications in the nuclear industry because of their high resistance to radiation and are candidates for potential ceramic waste host [65–68] and have also been suggested for use within new types of nuclear reactors. Ceramic-glass composites made from Mg-Al spinels and borosilicate glass can be used for ceramic boards for large scale integrated circuits used at high temperatures [69].

The presence of Al(III) ions is believed to inhibit formation of the SiO_2 polymorph cristobalite, which degrades the mechanical and electrical properties of these specialized ceramics. Glass-spinel ceramics have the chemical and thermal resistance usually associated with aluminates and also low thermal expansion and a low dielectric constant. If there is formation of cristobalite in these types of composites, then the thermal expansion can be uneven. Temperature-dependent formation of additional SiO_2 polymorphs can lead to micro-fracturing and mechanical degradation. Decreased ceramic contents of composites improve signal transfer by further lowering of the dielectric constant and so ideally the material will have a balance of spinel and glass optimized for the improved electrical properties and minimal cristobalite formation.

11   MgO–Al$_2$O$_3$ System

The binary phase diagram for MgO–Al$_2$O$_3$ is simpler than that for the CaO–Al$_2$O$_3$ system (Fig. 2). There is only one stable intermediate compound that of the spinel phase (Mg$_2$AlO$_4$) [60]. Spinel melts at 2,105°C, but there is a eutectic at 1,995°C and a limited solid solution between stoichiometric spinel and MgO (periclase), up to 6 wt% MgO, can be dissolved into the spinel structure without exsolution. This limited solid solution is an important property that is utilized in manufacture of spinels for use in reducing conditions [70].

The cubic spinel crystal structure (Fd3m) is a close-packed array of oxygen ions, which has the general form AB$_2$O$_4$. A is a divalent cation and B trivalent [60, 71].

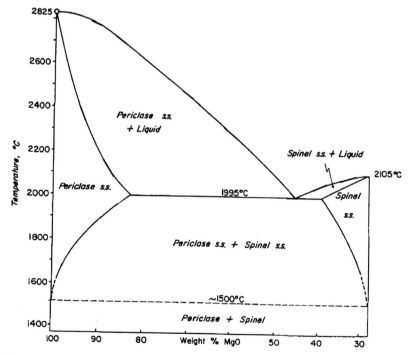

Fig. 2   The MgO–Al$_2$O$_3$ phase diagram [59]

The metal cations occupy two sites: divalent cations (A) are in tetrahedral coordination, while trivalent ions (B) occupy octahedral sites. The oxygen ions form a face-centered cubic close-packed arrangement and the unit cell consists of 32 oxygen ions, 8 divalent (A), and 16 trivalent ions (B) with dimensions of 0.80832 nm. There are a large number of natural forms of spinel structure, which include Cr_2O_3 and Fe_2O_3 forms. The lattice parameter A_o is 0.80832 nm, and in synthetic spinels, the limited solid solutions with both Al_2O_3 and MgO end-members are accommodated in this cubic structure, although there is slight increase in the lattice parameter [16].

There are two types of spinel, normal and inverted. Normal spinels have all the A ions in tetrahedral sites and all B ions in octahedral coordination. When the structures are inverted, the divalent A ions and half of the trivalent B ions are in the octahedral sites while the remaining B ions have tetrahedral coordination. Both normal and inverted spinels have the same cubic structure (space group Fd3m).

In high radiation fields, the spinel crystal structure has been shown to change. The structure, while still cubic, becomes disordered with a reduction in lattice parameter. The disordered "rock-salt" structure has a smaller unit cell reflecting the more random occupation of the octahedral sites by both trivalent and divalent ions. Increased radiation damage results in the formation of completely amorphous spinels. Radial distribution functions ($g(r)$) of these amorphous phases have Al–O and Mg–O radial distances that are different from equivalent crystalline phases. The Al–O distance in the amorphous form is reduced from Al–O of 0.194 nm in the crystalline phase to 0.18 nm in the amorphous phase, while the Mg–O distance is increased (0.19 nm in the crystal to 0.21 nm in the amorphous phase). Differences between the Al–O distances of crystalline and amorphous phases are a characteristic of both calcium and rare earth aluminates.

The $MgO–Cr_2O_3$ binary is closely related to the equivalent Al_2O_3 system. Here too the only stable compound is a spinel-structured phase $MgCr_2O_4$, which has a high melting point (2,350°C). The chrome-bearing ceramics have similar applications but have a significant drawback environmentally. There is a risk that chrome-bearing ceramics in furnace waste will interact and contaminate ground water. Cr[VI] ions leached from remnant refractory materials in wastes into ground water are a serious contaminant and have been linked to skin ulceration and carcinoma. $MgO–Al_2O_3$ ceramics are, therefore, much more desirable.

12 Synthesis of Magnesium Aluminates

As with many ceramics, $MgAl_2O_4$ spinels can be made by solid-state sintering of the component oxides MgO and Al_2O_3 [72]. Pure stoichiometric spinel ($MgAl_2O_4$) is made by solid-state reaction of high purity end-members at high temperatures. Starting materials are either oxides (Al_2O_3 and MgO) or carbonates ($MgCO_3$). The synthesis relies on solid-state reactions between the grains of starting material and so depends on the fineness of the powders used. An additional problem is the potential for Mg(II) to volatilize at high temperatures from the Mg-starting powder, which can lead to nonstoichiometric phases. In some instances this is desired, since more Al_2O_3-rich spinels are more stable under reducing atmospheres.

For some applications, better control on porosity is required and alternatives to solid-state synthesis methods have been sought requiring synthesis temperatures much lower than those used for the sintering route (1,600–1,800°C).

Chemical synthesis of $MgAl_2O_4$ spinels has been attempted using gibbsite ($Al(OH)_3$) and MgO precursors[73]. Spinel precursors are formed by coprecipitation and the resultant material is then calcined to produce spinel. The starting material gibbsite, which is a by-product of the Bayer process, is dissolved in a solution of HCl and HNO_3. MgO is added in a molar ratio 2:1 Al/Mg (i.e., stochiometric spinel). A precipitate is formed by adding NH_4OH to maintain a pH of 8.5–9.0. The precipitate is filtered and rinsed before calcining at temperatures of up to 1,400°C. Finally, nanophase spinel aggregates are formed with a reduced (83%) density.

Even greater control of the microstructure of spinels is achieved by joint crystallization of mixtures of magnesium and aluminum salts [74, 75]. The magnesium salt, magnesium nitrate hexahydrate ($Mg(NO_3)_2 \cdot 6H_2O$) can be used to form highly reactive spinel precursors by mixing in solution with different aluminum compounds. Vasilyeva and coworkers [74, 75] for example report synthesis of nano-phase spinels with porosity of up to 50% through use of aluminum nitrate monohydrate ($Al(NO_3)_3 \cdot 9H_2O$), aluminum isopropoxide ($Al((CH_3)_2CHO)_3$), and aluminum hydroxide (AlOOH, Boehmite). The stoichiometric mixtures of salts are dissolved in water and the pH is adjusted by the addition of nitric acid (HNO_3). The solutions are evaporated and then calcined at 250–900°C. The porosity is variable and depends on the aluminum compound used, and a combustible synthesis aid such as carbon can be added to further increase the porosity.

Sol–gel techniques have also been developed to make $MgAl_2O_4$ spinel [76]. In some applications [15, 77], such as filtration membranes for the food industry, spinels, which have greater chemical stability, are prepared on the surfaces of γ-Al_2O_3 nanoparticles. In this technique, boehmite, produced by sol–gel process, is used as a starting sol. In situ modification of the sol surface is achieved by adding $Mg(NO_3)_2$ and ethylene-dinitro-tetra-acetic acid (EDTA) to the aged boehmite sol, and polyvinyl acetate (PVA) solution and polyethylene glycol (PEG) is added to prevent defect formation. During calcining, at 550–850°C, magnesium oxide diffuses to the core and reacts with the alumina to form a spinel coat on the γ-Al_2O_3 particles.

A modified sol–gel method can also be used to make spinel directly [76]. Magnesium oxide is dispersed into an isopropanol solution of aluminum sec-butoxide. Water is added to the solution to promote alkoxide gelation and the slurry is evaporated to remove excess water and alcohol. The precursors are then dried and calcined at 300–800°C. In this case the formation of spinel is through reaction of nanophase MgO and Al_2O_3 in the spinel precursor during the calcining process.

13 Lithium Aluminates

Lithium aluminates have a potentially important role in the development of new types of nuclear reactors [78–81]. This role is a result of the nuclear reaction between the ^6Li isotope and neutrons $^6Li(n,\alpha)$, which results in a tritium (^3H) ion. The natural abundance of ^6Li is 7.5%, so ceramics can be made without any need for isotopic enrichment. The ^3H ions are the plasma fuel for fusion devices. The design of the ceramic requires a high mechanical and thermal stability so aluminates are often considered; because the operating conditions require diffusion of ^3H through pores, special synthesis conditions are required.

Lithium aluminates are also important in the development of molten carbonate fuel cells (MCFC) [82, 83]. In these fuel cells, a molten carbonate salt mixture is used as an electrolyte. These fuel cells operate through an anode reaction, which is a reaction between carbonate ions and hydrogen. A cathode reaction combines oxygen, CO_2, and electrons from the cathode to produce carbonate ions, which enter the electrolyte. These cells operate at temperatures of ~650°C and the electrolyte, which is usually lithium and potassium carbonate, is suspended in an inert matrix, which is usually a lithium aluminate.

14 Li$_2$O–Al$_2$O$_3$ System

As with the other aluminate systems, the binary Li_2O–Al_2O_3 is characterized by the two refractory oxide end-members, Li_2O (which melts at 1,000°C) and Al_2O_3. There are three stable compounds in the Li_2O–Al_2O_3 system: Li_5AlO_4, $LiAlO_2$, and $LiAl_5O_8$. The phases of most interest for materials application are the $LiAlO_2$ compounds that have α, β, and γ form (Fig. 3). The α-$LiAlO_2$ is orthorhombic and has the space group r3m, while the γ form has even lower (tetragonal) symmetry. Both Li and Al are in tetrahedral coordination in the γ phase. The γ phase can be produced irreversibly by sintering the α

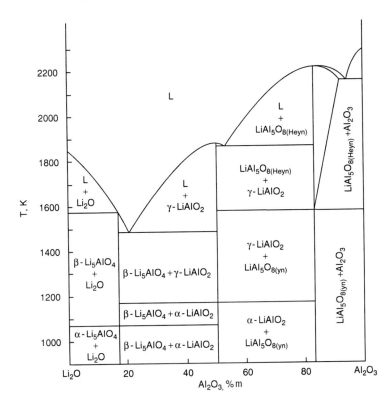

Fig. 3 The Li$_2$O–Al$_2$O$_3$ phase diagram [83]

phase at 1,350°C. The lattice parameters of $LiAlO_2$ match closely those of gallium nitrides, and lithium aluminates are used as a substrate for GaN epitaxial films.

15   Synthesis of Lithium Aluminates

The synthesis of lithium aluminates for tritium production requires formation of nanostructured phases. These can be made by solid-state reaction, by appropriate mixing of oxide powders [84] or by sol–gel methods [80, 85–87]. One technique is the peroxide route where γ-Al_2O_3 and $LiCO_3$ are dissolved in a peroxide (H_2O_2) solution. Evaporation of water and calcining the solid residue results in nanophase $LiAlO_2$.

Sol–gel synthesis of $LiAlO_2$ involves an alcohol–alkoxide route. Different alcohols and alkoxide combinations can be used. The alkoxide and alcohols are mixed and hydrolyzed by addition of pure water. The mixture is then gelled by heating to 60°C and is subjected to hydrothermal treatment in an autoclave. The morphology of the crystalline phases is dependent on the length of the alkoxy groups used in the alkoxide–alcohol mixing step. For example, rod-like crystals are produced with butoxide and propoxide mixtures.

16   Rare Earth Aluminates

Rare earth aluminates are extremely important as laser host materials. Most interest is in the system Y_2O_3–Al_2O_3, and of the three crystalline phases that are important in this system, the garnet phase (YAG, $Y_3Al_5O_{12}$) is the most important laser host. Laser hosts require highly transparent single crystals, and crystal growth studies of YAG were preformed at various laboratories in the 1950s and 1960s. YAG is optically isotropic and transparent from 300 nm to 4 µm and can accept trivalent laser activator ions. In 1964, the Bell Telephone Laboratories reported the lasing of YAG doped with Nd^{3+} [88]. Single crystals of YAG can be polished to form durable optical components of uniform refractive index and good thermal conductivity. Because of their robustness, YAG-based lasers have a wide variety of military, industrial, and medical applications.

Although Nd:YAG requires large, defect-free single crystals [89], polycrystalline ceramics are cheaper to manufacture and there is increasing use of YAG ceramics [90] as scintillators for radiation detection, for example Ce-doped YAG ceramics [91, 92]. In this case, the luminescence comes both from the activation of the Ce ion, with additional UV contribution, and from the YAG host itself.

Polycrystalline rare earth aluminates can also be used as advanced ceramic materials because of their refractory nature and chemical and mechanical durability. Y_2O_3–Al_2O_3 coatings are used in crystalline fibers, and Eu-doped Y-Al powders are used for phosphors and scintillation applications. Yttria and alumina are also used as additives for liquid phase synthesis of silicon nitride ceramics, often forming glassy coatings to the nitride phase and therefore having an important role in forming nitride-glass composites.

Because of the applications in laser and other optical devices, the Y_2O_3–Al_2O_3 system has been particularly well studied. There are three important crystal phases in the Y_2O_3–Al_2O_3 system: in addition to the cubic garnet phase, YAG ($Y_3Al_5O_{12}$), there is a perovskite phase (YAlO$_3$) and a monoclinic phase ($Y_4Al_2O_9$). Not all rare earth aluminate binaries have a stable garnet phase. Not all rare earth aluminate systems have a stable garnet phase. The stability of the garnet phase depends on the identity of the rare earth ion [93], for larger ionic radii (e.g., La(III)), the garnet phase is not stable and only pervoskites are present.

17 Y_2O_3–Al_2O_3 System

The Y_2O_3–Al_2O_3 phase diagram [11, 94, 95] has been studied extensively [96]. Large single crystals of YAG required for laser hosts are grown from the liquid phase by the Czochralski technique and a good knowledge of the phase equilibria is required to avoid formation of phases less suited for laser and scintillation applications [11, 97, 98].

There are five crystalline compounds in the Y_2O_3–Al_2O_3 binary system (Fig. 4). The two end-members have well-known crystal structures. Y_2O_3 is cubic with two YO_6 environments and α-Al_2O_3 is trigonal with one aluminum site of AlO_6 and one oxygen site of OAl_4. The crystalline compounds show two common aluminum coordination environments (^{IV}Al and ^{VI}Al) and a range of Y–O units (YO_6, YO_7, and YO_8). There is a large range of oxygen environments. The range in coordination environments for the three ions in the Y–O–Al system have been extensively studied by nuclear magnetic resonance spectroscopy (NMR) using the ^{27}Al, ^{89}Y, and ^{17}O nuclei [99].

The garnet phase (YAG, composition $Y_3Al_5O_{12}$) has a complicated crystal structure (Fig. 5) [17, 18, 99, 100]. The garnet phase is cubic, with 160 atoms per unit cell. There are two aluminum environments in YAG: one coordinated by four oxygen atoms (^{IV}Al) and one coordinated by six oxygen atoms (^{VI}Al). There is an unique yttrium site in YAG (YO_8) and a single, distinctive oxygen environment (OY_2A_{12}).

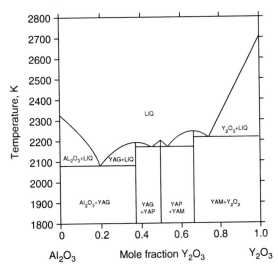

Fig. 4 The Y_2O_3–Al_2O_3 phase diagram [96–98]

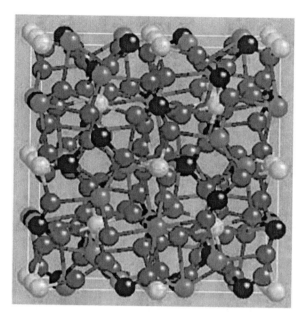

Fig. 5 The crystal structure of garnet (YAG)

The perovskite phase ($YAlO_3$) is orthorhombic and has a single type of ^{VI}Al-site (AlO_6) and a single YO_8 site. In contrast to the garnet phase, there are two oxygen sites: OY_2A_{12} and OY_3A_{12}. The monoclininc phase, YAM, has two different AlO_4 environments and four yttrium ions: two in YO_6 and two in YO_7. There are nine oxygen sites in YAM, four OY_3Al, two OY_2Al, two OY_4, and one OY_2A_{12}.

Thermodynamic data from solution calorimetry using molten lead borate [95, 101] for the perovskite and garnet phases have been combined with data for the YAM phase, and heat capacity data from adiabatic calorimetry and differential scanning calorimetry have been used to calculate the phase diagram for the binary system. This recent study shows unequivocally congruent melting of the perovskite phase and that it does not decompose to the YAM phase + liquid [95].

Studies of the liquid state of Y_2O_3–Al_2O_3 liquids close to YAG have revealed complex relations in the liquid state and the existence of a metastable eutectic at 23% Y_2O_3–77% Al_2O_3 [96, 98, 102] projected form the $YAlO_3$ composition and from α-Al_2O_3. The melting temperature at this eutectic composition is 1,702°C, considerably lower than that of the melting point of YAG itself (1,940°C). The presence of this eutectic has important implications for the nucleation behavior of YAG. Under some conditions the Y_2O_3–Al_2O_3 liquids can be deeply undercooled and form a eutectic mixture of α-Al_2O_3 and YAG. This is a reflection of the difficulty in forming YAG nuclei [103], implying major differences in the local structure of the YAG-liquids and the YAG crystal phase. YAG will only form if composition are heated to a temperature of ~50°C or less above the liquid temperature [98, 104], if heated to higher temperature the liquid can be supercooled considerably below the liquid temperature and depending on conditions form a mixed ceramic or a glass (if under containerless conditions [105–107]).

Undercooled Y_2O_3–Al_2O_3 liquids are notable in that they show an unusual form of transition. This is a so-called polyamorphic transition, which is a transition from a high density to low density amorphous phase without a change in composition [108]. This transition has been reproduced by several groups and the resultant samples comprise two glassy phases, which are amorphous. It is also to be noted that the composite samples do not consist of glass and nanocrystalline regions, as suggested by other groups [109].

18   Synthesis of YAG and Other Rare Earth Aluminates

A requirement for making YAG crystals is a homogeneous, high-purity starting material for use in the Czochralski technique. This means that the application of mechanical mixing and solid-state diffusion techniques are limited. Accordingly, a variety of synthesis techniques have been developed, most of which are sol–gel based.

The sol–gel technique is attractive because it allows for molecular mixing of constituents and results in chemical homogeneity. Typically calcining occurs at temperatures well below those required for solid-state synthesis resulting in amorphous nanocrystalline YAG samples. The citrate-based technique is commonly used for YAG synthesis [14, 110, 111]. The starting materials are aluminum and yttrium nitrates, which are soluble in water. Citric acid is added to the aqueous solution of the stoichiometric mixtures of nitrates (3:5 Y to A for YAG) to act as a chelating agent that is to stabilize the solution against hydrolysis or condensation. Ammonia is added to reduce excess acidity. The sol–gel process requires formation of an organic polymer framework, independent of the mineral species in solution. In the citrate technique, the polymer framework is based on acrylamide, which is easily soluble in water. Polymerization is initiated by free radicals and radical transfer agents. Transparent gels are obtained by heating to temperatures of ~80°C. The organic components and water are removed by placing the gel in a ventilated furnace and heating. Temperatures of 800°C result in amorphous YAG, while higher temperatures result in nanocrystalline YAG. Doping of YAG with other rare earth elements (e.g., Nd, Eu, or Tb) for phosphor or laser applications can be achieved by addition of the appropriate nitrate at the solution stage.

A translucent solution is produced by stoichiometric mixtures of aluminum isopropoxide and yttrium acetate in butanediol and by adding glycol instead of water in the autoclave (the so-called glycothermal technique) heated to 300°C. Adding ammonium hydroxide solution causes particles of YAG to precipitate.

There are some alternatives to sol–gel based on combustion synthesis [112–114]. The motivation of combustion synthesis is similar to that for sol–gel; a homogenous high-purity product. Coprecipitation of organo-metallic products is followed by addition of a fuel, such as urea or glycerin. When heated, combustion occurs causing localized formation of YAG.

19 Y_2O_3–Al_2O_3 Glasses

Liquids in the Y_2O_3–Al_2O_3 system do not form glasses very easily, a feature that is not just coincidental with the polyamorphic transition [115–117]. Poor glass-forming ability is a feature of "fragile" liquids and the polyamorphic transition in Y_2O_3–Al_2O_3

liquids is interpreted as a transition from a fragile to strong liquid. This has important implications for materials applications of YAG and closely-related glasses.

Y_2O_3–Al_2O_3 glasses have high mechanical strength and desirable optical properties so there are potential commercial applications for YAG fibers in composite materials and optical devices. Glasses can be made from YAG using container-less levitation [13, 105, 107, 118] with fibers drawn from the beads. The lengths of the fibers are limited and this is most likely due to the changes in rheology that occur as the transition is intercepted. Furthermore, when there is formation of a second glassy phase, different properties occur leading to a "necking" and breaking of the fiber.

The unusual behavior of Y_2O_3–Al_2O_3 glasses had resulted in extensive calorimetric and diffraction studies. These suggest that the glass structure is very different from the crystalline phases [119]. Although the crystalline phases are dominated by octahedral aluminum (^{VI}Al), this short range order is reduced in the amorphous phases and the mean Al–O coordination number is close to 4 (4.15) [119]. A similar change in Y–O coordination is reported and the glass phases are not simply disordered forms of the crystalline equivalents. This observation is consistent with the fragility of Y_2O_3–Al_2O_3 liquids, and indeed molecular dynamics simulations (which are at much higher temperatures than the fictive temperatures of the glass) suggest that the Al–O coordination number increases with temperature, as further suggested by studies of levitated liquids. A structural model for a single phase (high density) Y_2O_3–Al_2O_3 glass is shown in Fig. 6.

Glasses of $Y_3Al_5O_{12}$ composition can be made by levitation techniques [106] and other glass compositions can be made at compositions close to that of the metastable eutectic (Fig. 4) using a Xe-arc image furnace. The precursors are made via sol–gel route [115].

Commercial products are being developed from single phase aluminate glasses [107, 120]. These "REAL™ Glass" materials were first made using levitation melting. Subsequently, formulations that can be cast from melts formed in platinum crucibles have been developed. The glasses are hard, strong, and environmentally stable and

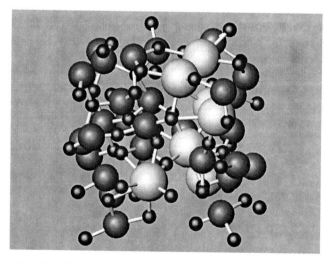

Fig. 6 Structural model of 20% Y_2O_3–80% Al_2O_3 glass

Fig. 7 A 1-cm diameter, 2-mm thick optical window made from REAL™ glass

they can be doped with high concentrations of rare earth ions for use in optical devices. Initial commercial applications are as an alternative to sapphire for use in infrared windows and optics that transmit to a wavelength of approximately 5 μm (Fig. 7). Rare earth aluminate glasses are also important medically [121]. Yttrium in rare earth aluminates can be activated by neutrons to form short-lived radioactive isotopes (^{90}Y) for cancer therapy.

20 Conclusions

Aluminate glasses and ceramics include a range of compositions and crystal structures. Most of the important physical properties of aluminates are similar to those of Al_2O_3. These properties include high melting point and high mechanical strength. Aluminate ceramics are frequently binary systems and have intermediate compounds that, while retaining the relatively high melting point, can be easily synthesized.

Aluminate compositions include calcium aluminate cements, which have high chemical resistance, especially to sulfate, and is also used in refractory applications where ordinary Portland cements would be unsuitable. These same cements are used in bioceramic applications. The bioceramic applications reflect both the high mechanical strength of the calcium aluminate cements and also the "biocompatibility" of Ca-bearing phases, which bond well with, for example, bone.

Other binary aluminates include magnesium spinels that are used extensively as castable refractory ceramics. Lithium aluminates are used, potentially, in fuel cells and as materials for new types of nuclear reactors. Again, these applications reflect the refractory nature of aluminates and their chemical resistance.

Rare earth aluminates are also important commercially as ceramics and ceramic composites for scintillation applications. The importance of the optical properties of rare earth aluminates is underscored by the used of Nd-doped YAG as a laser host.

Synthesis of aluminates is in the most part a solid-state process using purified components and requiring high temperatures. Sol–gel techniques are also used, since this is a lower temperature route and also because in many applications grain size and porosity need to be controlled.

Glasses can be formed from aluminates, but the glass-forming ability is poor. This reflects the fragility of aluminate liquids which, in Y_2O_3–Al_2O_3 systems leads to anomalous thermodynamic properties. As a result, exotic techniques are used to make aluminate glasses, most importantly container-less levitation.

Acknowledgements Dr J. K. R Weber from Materials Development Inc. kindly commented on an earlier draft of this manuscript and also provided details of the REAL™ glasses and Fig. 7. I also thank Dr. J. F. Shackelford for his support and encouragement to ensure completion of this chapter.

References

1. K.L. Scrivener, J.L. Cabiron, and R. Letourneux, High-performance concretes from calcium aluminate cements. Cement Concrete Res. **29**(8), 1215–1223 (1999).
2. K.L. Scrivener, Historical and present day applcaiitons of calcium aluminate cements, in *Calcium and Calcium Aluminate Cements*, R.J. Mangabhai and F.P. Glasser (eds.), London IOM Communications, (2001).
3. Bensted, J., High alumina cement – present state of knowledge. *Zement-Kalk-Gips*, **46**(9), 560–566 (1993).
4. Y. Liu et al., Behaviour of composite ca/p bioceramics in simulated body fluid. *Mate. Sci. Technol* **14**(6), 533–537 (1998).
5. H. Engqvist et al., Transmittance of a bioceramic dental restorative material based on calcium aluminate. *J. Biomed. Mater. Res. B-Appl. Biomater.* **69**B(1), 94–98 (2004).
6. H. Engqvist et al., Chemical and biological integration of a mouldable bioactive ceramic material capable of forming apatite in vivo in teeth. *Biomaterials*, **25**(14), 2781–2787 (2004).
7. J. Loof et al., Mechanical properties of a permanent dental restorative material based on calcium aluminate. *J. Mater. Sci.-Mater. Med.* **14**(12), 1033–1037 (2003).
8. S.H. Oh et al., Preparation of calcium aluminate cement for hard tissue repair: Effects of lithium fluoride and maleic acid on setting behavior, compressive strength, and biocompatibility. *J. Biomed. Mater. Res.* **62**(4), 593–599 (2002).
9. M.K. Cinibulk et al., Porous yttrium aluminum garnet fiber coatings for oxide composites. *J. Am. Ceram. Soc.* **85**(11), 2703–2710 (2002).
10. M.K. Cinibulk, K.A. Keller, and T.I. Mah, Effect of yttrium aluminum garnet additions on alumina-fiber-reinforced porous-alumina-matrix composites. *J. Am. Ceram. Soc.* **87**(5), 881–887 (2004).
11. L.V. Soboleva and A.P. Chirkin, Y2O3-Al2O3-Nd2O3 phase diagram and the growth of (Y,Nd)(3)Al5O12 single crystals. *Crystallog. Rep.* **48**(5), 883–887 (2003).
12. E.M. Nunes et al., A volume radiation heat transfer model for Czochralski crystal growth processes. *J. Cryst. Growth* **236**(4), 596–608 (2002).
13. J.K.R. Weber et al., Aero-acoustic levitation - a method for containerless liquid-phase processing at high temperatures. *Rev. Sci. Instrum.* **65**(2), 456–465 (1994).

14. Y.H. Zhou et al., Preparation of Y3Al5O12: Eu phosphors by citric-gel method and their luminescent properties. *Opt. Mater.* 2002. **20**(1), 13–20 (2002).

15. X.L. Pan et al., Mesoporous spinel MgAl2O4 prepared by in situ modification of boehmite sol particle surface: I Synthesis and characterization of the unsupported membranes. Colloids Surf. A-Physicochem. *Eng. Asp.* **179**(2–3), 163–169 (2001).

16. K.E. Sickafus, J.M. Wills, and N.W. Grimes, Structure of spinel. *J. Am. Ceram. Soc.* **82**(12), 3279–3292 (1999).

17. C.J. Howard, B.J. Kennedy, and B.C. Chakoumakos, Neutron powder diffraction study of rhombohedral rare-earth aluminates and the rhombohedral to cubic phase transition. *J. Phys.-Conden. Matter.* **12**(4), 349–365 (2000).

18. L. Vasylechko et al., Crystal structure of GdFeO3-type rare earth gallates and aluminates. *J. Alloys Comp.* **291**(1–2), 57–65 (1999).

19. C.E. Johnson, K. Noda, and N. Roux, Ceramic breeder materials: status and needs. *J. Nucl. Mater.* **263**(Part A) 140–148 (1998).

20. M. Poulain, Advanced glasses. *Annales de Chimie-Science des Materiaux.* **28**(2), 87–94 (2003).

21. W.A. King, and J.E. Shelby, Strontium calcium aluminate glasses. *Phys. Chem. Glasses* **37**(1), 1–3 (1996).

22. A.G. Holterhoff, Calcium aluminate cements. *Am. Ceram. Soc. Bull.* **73**(6), 90–91 (1994).

23. A.G. Holterhoff, Calcium aluminate cements. *Am. Ceram. Soc. Bull.* **74**(6), 117–118 (1995).

24. A.K. Chatterjee, An update on the binary aluminates appearing in aluminous cements, in *Calcium and Calcium Aluminate Cements*, R.J. Mangabhai and F.P. Glasser (eds.), London, IOM Communications, (2001).

25. D.A. Jerebtsov and G.G. Mikhailov, Phase diagram of CaO-Al2O3 system. *Ceram. Int.* **27**(1) 25–28 (2001).

26. B. Hallstedt, Assessment of the Cao-Al2O3 system. *J. Am. Ceram. Soc.* **73**(1), 15–23 (1990).

27. H. Pollmann, Mineralogy and crystal chemistry of calcium aluminate cement, in *Calcium and Calcium Aluminate Cements*, R.J. Mangabhai and F.P. Glasser, (eds.), London, IOM Communications (2001).

28. D.A. Brosnan and H.D. Leigh, Rehydration of castable refractories. *Can. Ceram. Q.-J. Can. Ceram. Soc.* **64**(2), 122–126 (1995).

29. W.E. Lee. et al., Castable refractory concretes. *Int. Mater. Rev* **46**(3), 145–167 (2001).

30. K. Konradsson, and J.W.V. van Dijken, Effect of a novel ceramic filling material on plaque formation and marginal gingiva. *Acta Odontol Scand.* **60**(6), 370–374 (2002).

31. S.J. Kalita, et al., Porous calcium aluminate ceramics for bone-graft applications. *J. Mater. Res.* **17**(12), 3042–3049 (2002).

32. I.N. Chakraborty, and A.K. Chattopadhyay, Manufacture of high alumina cement, an indian experience, in *Calcium and Calcium Aluminate Cements*, R.J. Mangabhai and F.P. Glasser (eds.), London, IOM Communications, 2001.

33. D.A. Fumo, M.R. Morelli, and A.M. Segadaes, Combustion synthesis of calcium aluminates. *Mater. Res. Bull.* **31**(10), 1243–1255 (1996).

34. Singh, V.K., Sintering of calcium aluminate mixes. *Br. Ceram. Trans.* **98**(4), 187–191 (1999).

35. M. Uberoi, and S.H. Risbud, Processing of Amorphous Calcium Aluminate Powders at Less-Than 900-Degrees-C. *J. Am. Ceram. Soc.* **73**(6), 1768–1770 (1990).

36. A. Varma, and A.S. Mukasyan, Combustion synthesis of advanced materials: Fundamentals and applications. Korean *J. Chem. Eng.* **21**(2), 527–536 (2004).

37. T. Aitasalo, et al., EU2+ doped calcium aluminates prepared by alternative low temperature routes. *Opt. Mater.* **26**(2), 113–116 (2004).

38. J.A. Sampaio, and S. Gama, EXAFS investigation of local structure of Er3+ and Yb3+ in low-silica calcium aluminate glasses - art. no. 104203. *Phys. Rev. B* **6910**(10), 4203 (2004).

39. P.F. Paradis, et al., Contactless density measurement of liquid Nd-doped 50%CaO-50%Al2O3. *J. Am. Ceram. Soc.* **86**(12), 2234–2236 (2003).

40. C.J. Benmore, et al., A neutron and x-ray diffraction study of calcium aluminate glasses. *J. Phys. Conden Matter* **15**(31), S2413–S2423 (2003).

41. M.S.F. Da Rocha, et al., Radiation-induced defects in calcium aluminate glasses. Radiat. Eff. Defects Solids **158**(1–6), 363–368 (2003).

42. W.J. Chung and J. Heo, Energy transfer process for the glue up-conversion in calcium aluminate glasses doped with Tm3+ and Nd3+. *J. Am. Ceram. Soc.* **84**(2), 348–352 (2001).

43. D.F. de Sousa et al., Energy transfer and the 2.8-mu m emission of Er3+- and Yb3+-doped low silica content calcium aluminate glasses. *Phys. Rev. B* **62**(5), 3176–3180 (2000).

44. W.Y. Li, and B.S. Mitchell, Nucleation and crystallization in calcium aluminate glasses. *J. Non-Cryst. Solids* **255**(2–3), 199–207 (1999).

45. P.F. McMillan, et al., A structural investigation of cao-al2o2 glasses via al-27 mas-nmr. *J. Non-Cryst. Solids* **195**(3), 261–271 (1996).

46. E.V. Uhlmann, et al., Spectroscopic properties of rare-earth-doped calcium-aluminate-based glasses. *J. Non Cryst Solids.* **178**, 15–22 (1994).

47. C.A. Angell, Glass forming liquids with microscopic to macroscopic two-state complexity. *Progress Theor. Phys. Suppl.* **126**, 1–8 (1997).

48. A.C. Hannon, and J.M. Parker, The structure of aluminate glasses by neutron diffraction. *J. Non-Cryst.* **274**(1–3), 102–109 (2000).

49. J.K.R. Weber, et al., Novel synthesis of calcium oxide-aluminum oxide glasses. *Japanese J. App. Phys. 1-Regul. Pap. Short Notes Rev Pap.* **41**(5A), 3029–3030 (2002).

50. S. Mukhopadhyay, et al., In situ spinel bonded refractory castable in relation to co-precipitation and sol-gel derived spinel forming agents. *Ceram. Int.* **29**(8), 857–868 (2003).

51. C.J. Ting, and H.Y. Lu, Hot-pressing of magnesium aluminate spinel - I. Kinetics and densification mechanism. *Acta Mater.* **47**(3), 817–830 (1999).

52. C.J. Ting, and H.Y. Lu, Hot-pressing of magnesium aluminate spinel - II. Microstructure development. *Acta Mater.* **47**(3), 831–840 (1999).

53. M.W. Vance, et al., Influence of spinel additives on high-alumina spinel castables. *Am. Ceram. Soc. Bull.* **73**(11), 70–74 (1994).

54. J.W. Lee, and J.G. Duh, High-temperature MgO-C-Al refractories-metal reactions in high-aluminum-content alloy steels. *J. Mater. Res.* **18**(8), 1950–1959 (2003).

55. A. Ghosh, et al., *Effect of spinel content on the properties of magnesia-spinel composite refractory. J. Eur. Ceram. Soc.* **24**(7), 2079–2085 (2004).

56. K. Goto, B.B. Argent, and W.E. Lee, Corrosion of mgo-mgal2o4 spinel refractory bricks by calcium aluminosilicate slag. *J. Am. Ceram. Soc.* **80**(2), 461–471 (1997).

57. A.H. De Aza, et al., Corrosion of a high alumina concrete with synthetic spinel addition by ladle slag. *Boletin de la Sociedad Espanola de Ceramica y Vidrio* **42**(6), 375–378 (2003).

58. B. Hallstedt, The magnesium-oxygen system. *Calphad-Comp. Coupling Phase Diagrams Thermochem.* **17**(3), 281–286 (1993).

59. B. Hallstedt, Thermodynamic assessment of the system Mgo-Al2O3. *J. Am. Ceram. Soc.* **75**(6), 1497–1507 (1992).

60. H.S.C. Oneill, and A. Navrotsky, Cation distributions and thermodynamic properties of binary spinel solid-solutions. *Am. Mineral.* **69**(7–8), 733–753 (1984).

61. F.C. Klaassen, et al., Post irradiation examination of irradiated americium oxide and uranium dioxide in magnesium aluminate spinel. *J. Nucl. Mater.* **319**, 108–117 (2003).

62. G.P. Pells, Radiation effects in ceramics. *MRS Bulle.* **22**(4), 22–28 (1997).

63. I. Ganesh, et al., An efficient MgAl2O4 spinel additive for improved slag erosion and penetration resistance of high-Al2O3 and MgO-C refractories. *Ceram. Int.* **28**(3), 245–253 (2002).

64. V.T. Gritsyna, et al., Neutron irradiation effects in magnesium-aluminate spinel doped with transition metals. *J. Nucl. Mater.* **283**(Part B), 927–931 (2000).

65. S.E. Enescu, et al., High-temperature annealing behavior of ion-implanted spinel single crystals. *J. Mater. Res.* **19**(12), 3463–3473 (2004).

66. Y.W. Lee, et al., Study on the mechanical properties and thermal conductivity of silicon carbide-, zirconia- and magnesia aluminate-based simulated inert matrix nuclear fuel materials after cyclic thermal shock. *J. Nucl. Mater.* **319**, 15–23 (2003).

67. T. Wiss, and H. Matzke, Heavy ion induced damage in MgAl2O4, an inert matrix candidate for the transmutation of minor actinides. *Radiat. Meas.* **31**(1–6), 507–514 (1999).

68. K. Yasuda, C. Kinoshita, and R. Morisaki, Role of irradiation spectrum in the microstructural evolution of magnesium aluminate spinel. *Philos. Mag. A-Phys. Condens.* Matter Struct. Defects Mechani. Properties, **78**(3), 583–598 (1998).

69. A.A. El-Kheshen, and M.F. Zawrah, Sinterability, microstructure and properties of glass/ceramic composites. *Ceram. Int.* **29**(3), 251–257 (2003).

70. A.M. Alper, et al., The system MgO-MgAl2O4. *J. Am. Ceram. Soc.* **45**(6), 263–268 (1962).

71. M. Ishimaru, et al., Atomistic structures of metastable and amorphous phases in ion-irradiated magnesium aluminate spinel. *J. Phys.-Condens. Matter* **14**(6), 1237–1247 (2002).

72. R.E. Carter, Mechanism of solid state reaction between magnesium oxide and aluminum oxide and between magnesium oxide and ferric oxide. *J. Am. Ceram. Soc.* **44**(3), 116–120 (1960).

73. H. Reveron, et al., Chemical synthesis and thermal evolution of $MgAl2O4$ spinel precursor prepared from industrial gibbsite and magnesia powder. *Mater. Lett.* **56**(1–2), 97–101 (2002).

74. E.A. Vasil'eva, et al., A porous ceramic based on aluminomagnesium spinel. *Russ. J. Appl. Chem.* **75**(6), 878–882 (2002).

75. E.A. Vasil'eva, et al., Specific features of the synthesis of porous materials based on a magnesium-aluminum spinel. *Glass Phys. and Chem.* **29**(5), 490–493 (2003).

76. F. Oksuzomer, et al., Preparation of $MgAl2O4$ by modified sol-gel method, in Euro Ceramics Viii, Parts 1–3. 2004, pp. 367–370.

77. Y.X. Pan, M.M. Wu, and Q. Su, Comparative investigation on synthesis and photoluminescence of YAG: Ce phosphor. *Mater. Sci. Eng. B-Solid State Mater. Adv. Technol.* **106**(3), 251–256 (2004).

78. L.M. Carrera, et al., Tritium recovery from nanostructured $LiAlO2$. *J. Nucl. Mater.* **299**(3), 242–249 (2001).

79. C.E. Johnson, K. Noda, and N. Roux, Ceramic breeder materials: status and needs. *J. Nucl. Mater.* **263**, 140–148 (1998).

80. O. Renoult, et al., Sol-gel lithium aluminate ceramics and tritium extraction mechanisms. *J. Nucl. Mater.* **219**, 233–239 (1995).

81. T. Kawagoe, et al., Surface inventory of tritium on $Li2TiO3$. *J. Nucl. Mater.* **297**(1), 27–34 (2001).

82. V.S. Batra, et al., Development of alpha lithium aluminate matrix for molten carbonate fuel cell. *J. Power Sources* **112**(1), 322–325 (2002).

83. K. Nakagawa, et al., Allotropic phase transformation of lithium aluminate in MCFC electrolyte plates. *Denki Kagaku* **65**(3), 231–235 (1997).

84. S. Sokolov, and A. Stein, Preparation and characterization of macroporous gamma-$LiAlO2$. *Mater. Lett.* **57**(22–23), 3593–3597 (2003).

85. M.A. Valenzuela, et al., Solvent effect on the sol-gel synthesis of lithium aluminate. *Mate. Lett.* **47**(4–5), 252–257 (2001).

86. M.A. Valenzuela, et al., Sol-gel synthesis of lithium aluminate. *J. Am. Ceram. Soc.* **79**(2), 455–460 (1996).

87. F. Oksuzomer, et al., Effect of solvents on the preparation of lithium aluminate by sol-gel method. *Mater. Res. Bulle.* **39**(4–5), 715–724 (2004).

88. J.E. Geusic, H.M. Marcos, and V.U. L.G., Laser oscillations in Nd-doped yttrium aluminum, yttrium gallium, and gadolinium garnets. *App. Phys. Lett.* **4**, 182–184 (1964).

89. E. Zych, C. Brecher, and J. Glodo, Kinetics of cerium emission in a YAG: Ce single crystal: the role of traps. *J. Phys. Condens. Matter.* **12**(8), 1947–1958 (2000).

90. J.R. Lu, et al., Neodymium doped yttrium aluminum garnet ($Y3Al5O12$) nanocrystalline ceramics – a new generation of solid state laser and optical materials. *J. Alloys Compd.* **341**(1–2), 220–225 (2002).

91. E. Zych, and C. Brecher, Temperature dependence of Ce-emission kinetics in YAG: Ce optical ceramic. *J. Alloys Compd.* **300**, 495–499 (2000).

92. E. Zych, et al., Luminescence properties of ce-activated yag optical ceramic scintillator materials. *J. Luminescence* **75**(3), 193–203 (1997).

93. Y. Kanke, and A. Navrotsky, A calorimetric study of the lanthanide aluminum oxides and the lanthanide gallium oxides: stability of the perovskites and the garnets. *J. Solid State Chem.* **141**(2), 424–436 (1998).

94. M. Medraj, et al., High temperature neutron diffraction study of the $Al2O3$-$Y2O3$ system. *J. Eur. Ceram. Soc.* **26**(16), 3515–3524 (2006).

95. O. Fabrichnaya, et al., The assessment of thermodynamic parameters in the $Al2O3$-$Y2O3$ system and phase relations in the Y-Al-O system. *Scand. J. Metall.* **30**(3), 175–183 (2001).

96. J.L. Caslavsky, and D. Viechnicki, Phase-equilibria studies in the ternary-system $Al2O3$/$Y2O3$/$Nd2O3$ using odta. *Am. Ceram. Soc. Bulle.* **61**(8), 808–808 (1982).

97. J.L. Caslavsky, and D. Viechnicki, Melt Growth of Nd - $Y3al5o12$ (Nd-Yag) Using the Heat-Exchange Method (Hem). *J. Cryst. Growth.* **46**(5), 601–606 (1979).

98. J.L. Caslavsky, and D.J. Viechnicki, Melting behavior and metastability of yttrium aluminum garnet (Yag) and $YAlO3$ determined by optical differential thermal-analysis. *J. Mater. Sci.* **15**(7), 1709–1718 (1980).

99. P. Florian et al., A multi-nuclear multiple-field nuclear magnetic resonance study of the $Y2O3$-$Al2O3$ phase diagram. *J. Phys. Chem. B.* **105**(2), 379–391 (2001).

100. I. Zvereva et al., Complex aluminates RE2SrAl2O7 (RE = La, Nd, Sm-Ho): Cation ordering and stability of the double perovskite slab-rocksalt layer P-2/RS intergrowth. Solid State Sci. 5(2), 343–349 (2003).
101. O. Fabrichnaya et al., Phase equilibria and thermodynamics in the Y2O3-Al2O3-SO2-system. Zeitschrift fur Metallkunde 92(9), 1083–1097 (2001).
102. B. Cockayne, and B. Lent, A complexity in the solidification behavior of molten Y3Al5O12. J. Cryst. Growth. 46, 371–378 (1979).
103. B.R. Johnson, and W.M. Kriven, Crystallization kinetics of yttrium aluminum garnet (Y3Al5O12). J. Mater. Res. 16(6), 1795–1805 (2001).
104. M. Gervais et al., Crystallization of y3al5o12 garnet from deep undercooled melt effect of the al-ga substitution. Mater. Sci. Eng. B-Solid State Mater. Adv. Technol. 45(1–3), 108–113 (1997).
105. J.K.R. Weber et al., Growth and crystallization of YAG- and mullite-composition glass fibers. J. Eur. Ceram. Soc. 19(13–14), 2543–2550 (1999).
106. J.K.R. Weber et al., Structure of liquid Y3Al5O12 (YAG). Phys. Rev. Lett. 84(16), 3622–3625 (2000).
107. J.K.R. Weber et al., Glass formation and polyamorphism in rare-earth oxide-aluminum oxide compositions. J. Am. Ceram. Soc. 83(8), 1868–1872 (2000).
108. S. Aasland, and P.F. McMillan, Density-driven liquid-liquid phase separation in the system al2o3-y2o3. Nature 369(6482), 633–636 (1994).
109. K. Nagashio, and K. Kuribayashi, Spherical yttrium aluminum garnet embedded in a glass matrix. J. Am. Ceram. Soc. 85(9), 2353–2358 (2002).
110. A. Douy, Polyacrylamide gel: an efficient tool for easy synthesis of multicomponent oxide precursors of ceramics and glasses. Int. J. Inorg. Mater. 3(7), 699–707 (2001).
111. J.J. Zhang et al., Low-temperature synthesis of single-phase nanocrystalline YAG: Eu phosphor. J. Mater. Sci. Lett. 22(1), 13–14 (2003).
112. J. Marchal et al., Yttrium aluminum garnet nanopowders produced by liquid-feed flame spray pyrolysis (LF-FSP) of metalloorganic precursors. Chem. Mater. 16(5), 822–831 (2004).
113. S.D. Parukuttyamma et al., Yttrium aluminum garnet (YAG) films through a precursor plasma spraying technique. J. Am. Ceram. Soc. 84(8), 1906–1908 (2001).
114. S. Ramanthan et al., Processing and characterization of combustion synthesized YAG powders. Ceram. Int. 29(5), 477–484 (2003).
115. M.C. Wilding, and P.F. McMillan, Polyamorphic transitions in yttria-alumina liquids. J. Non Cryst. Solids 293, 357–365 (2001).
116. M.C. Wilding, P.F. McMillan, and A. Navrotsky, Calorimetric study of glasses and liquids in the polyamorphic system Y2O3-Al2O3. Phys. Chem. Glasses 43(6), 306–312 (2003).
117. M.C. Wilding, P.F. McMillan, and A. Navrotsky, Thermodynamic and structural aspects of the polyamorphic transition in yttrium and other rare-earth aluminate liquids. Phys. Stat. Mech. Appl. 314(1–4), 379–390 (2002).
118. J.K.R. Weber et al., Glass fibres of pure and erkium- or neodymium-doped yttria-alumina compositions. Nature 393(6687), 769–771 (1998).
119. M.C. Wilding, C.J. Benmore, and P.F. McMillan, A neutron diffraction study of yttrium- and lanthanum-aluminate glasses. J. Non-Cryst. Solids 297(2–3), 143–155 (2002).
120. J.K.R. Weber, US patent 6,482,758; Single phase rare earth oxide aluminum oxide glasses. 2002: US.
121. T.E. Day, and D.E. Day, Manufacturing RadSpheres. Am. Ceram. Soc. Bull. 83(8), 21–21 (2004).

Chapter 5
Quartz and Silicas

Lilian P. Davila, Subhash H. Risbud, and James F. Shackelford

Abstract Silica is the most ubiquitous mineral in the earth's crust, existing in a wide variety of crystalline and noncrystalline forms due to the flexibility of the linkage among SiO_4 tetrahedra. The thermodynamically stable, room temperature form of silica is quartz, which is itself a widely available mineral and ingredient in many commercial ceramics and glasses. In addition to historically abundant raw material sources, crystalline and noncrystalline silicas can be produced by a wide range of synthetic routes. For example, synthetic quartz can be produced by hydrothermal growth in an autoclave, and synthetic vitreous silica can be produced from silicon tetrachloride by oxidation or hydrolysis in a methane–oxygen flame. Pure silicas serve as model systems in the study of ceramics and glasses, but at the same time, are used in a wide and steadily increasing variety of sophisticated technological applications, from piezoelectric crystals to optical fibers to waveguides in femtosecond lasers. Increased understanding of these ubiquitous materials is aided by improved experimental tools such as new neutron scattering facilities and increasingly sophisticated computer simulation methods.

1 Introduction and Historic Overview

Quartz and the silicas are composed of silicon and oxygen, the two most ubiquitous elements in the earth's crust [1] (Fig. 1). The widespread presence of the various forms of SiO_2 in common geological materials is a manifestation of this fact. Along this line, many common geological silicates (SiO_2-based materials such as rocks, clays, and sand) hold a detailed historical record of high-pressure and elevated temperature conditions with significant implications in materials science, engineering, geology, planetary science, and physics [2]. As a result, a discussion of pressure-related structure and properties will be included in this chapter.

Silica (SiO_2) is the most important and versatile ceramic compound of MX_2 stoichiometry. As noted above, it is widely available in raw materials in the earth's surface, and silica is a fundamental constituent of a wide range of ceramic products and glasses;

J.F. Shackelford and R.H. Doremus (eds.), *Ceramic and Glass Materials:*
Structure, Properties and Processing.
© Springer 2008

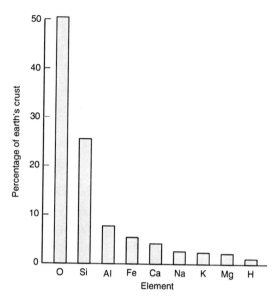

Fig. 1 The relative abundance of elements in the earth's crust illustrates the common availability of quartz and the silicas [1]

its properties permit it to be used in high-temperature and corrosive environments and as abrasives, refractory materials, fillers in paints, and optical components.

Vitreous silica (high purity SiO_2) is a technologically important amorphous material used in a myriad of applications including gas transport systems, laser optics, fiber optics, waveguides, electronics, vacuum systems, and furnace windows. During service, glass may experience elevated conditions of pressures and temperatures that can alter its properties. For instance, a vitreous silica lens may undergo drastic structural changes if pressure and temperature vary greatly in laser optics components. On the other hand, vitreous silica may undergo beneficial structural modifications under controlled conditions, e.g. during waveguide fabrication when femtosecond lasers are applied to induce a desired index of refraction in this glass [3].

2 The Structural Forms of Quartz and Other Silicas

Except for water, silica is the most extensively studied MX_2 compound. One of the challenges in studying silica is its complex set of structures. Silica has several common polymorphs under different conditions of temperature [1] and pressure [4], as seen in Figs. 2 and 3. For instance, cristobalite is the crystalline silica polymorph at atmospheric pressure above 1,470°C. It is built on an fcc lattice with 24 ions per unit cell. This structure is, in fact, the simplest form of silica. In addition to five polymorphs (quartz, coesite, stishovite, cristobalite, tridymite) that have thermodynamic stability fields, a large and increasing number of metastable polymorphs have been synthesized. These include vitreous silica, clathrasils, and zeolites [2]. Except for stishovite, all these structures are based on frameworks of

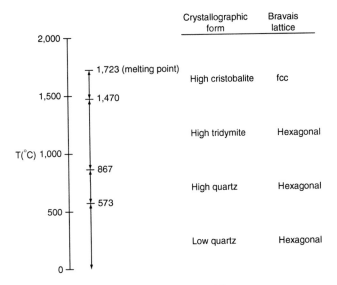

Fig. 2 Principal silica polymorphs at atmospheric pressure [1]

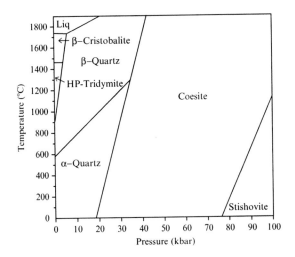

Fig. 3 Phase diagram for the SiO_2 system [4]

SiO_4 tetrahedra. These silica structures have been determined mainly by X-ray and neutron diffraction methods and, more recently, by Si and Al magic angle spinning solid-state NMR studies.

The various framework silica structures arise from the different ways that the $(SiO_4)^{4-}$ tetrahedra are linked into 1-, 2-, and 3-dimensional arrangements. Although the basic tetrahedra are present in most silica structures, the connectivity varies widely.

Both ionic and covalent natures of the Si–O bond contribute to the preference for $(SiO_4)^{4-}$ tetrahedron formation in both crystalline and glassy silicas. In addition, each O anion is coordinated by two Si cations, corresponding to corner sharing of the oxide tetrahedra, preventing the close-packing of anion layers and resulting in relatively open structures [5].

2.1 Silica Polymorphs

The name quartz comes from the German word "quarz," of uncertain origin. Quartz and the other main polymorphs of silica are related in the phase diagram [4] shown in Fig. 3. Under ambient conditions, α-quartz is the thermodynamically favored polymorph of silica. At 573°C, α-quartz is transformed into β-quartz, generally similar in structure but with less distortion. This thermal transformation preserves the optical activity of quartz. Heating quartz to 867°C leads to the transformation of β-quartz into β-tridymite, involving the breaking of Si–O bonds to allow the oxygen tetrahedra to rearrange themselves into a simpler, more open hexagonal structure of lower density. The quartz–tridymite transformation involves a high activation energy process that results in loss of the optical activity of quartz. Heating of β-tridymite to 1,470°C gives β-cristobalite that resembles the structure of diamond with silicon atoms in the diamond carbon positions and an oxygen atom midway between each pair of silicon atoms. Further heating of cristobalite results in melting at 1,723°C. A silica melt is easily transformed into vitreous silica by slow cooling, resulting in a loss of long-range order but retaining the short-range order of the silica tetrahedron.

In the last ten years, at least a dozen polymorphs of pure SiO_2 have been reported [6]. Stishovite, another form of silica obtained at high temperatures and pressures, has, rather than a tetrahedral-based geometry, a rutile (TiO_2) structure in which each Si atom is bonded to six O atoms and each O atom bridges three Si atoms [6]. Stishovite (found in Meteor Crater, Arizona) is more dense and chemically more inert than normal silica but reverts to amorphous silica upon heating.

The distinction among polymorphs other than stishovite arises from the different arrangements of connected tetrahedra. Important examples are quartz and cristobalite. The structures of these polymorphs are relatively complicated. These structures are also relatively open, as corner sharing of oxide tetrahedra prevents the close-packing of anion layers as found in the fcc- and hcp-based oxides [5]. One consequence is that these crystalline structures have low densities, e.g., quartz has a density of 2.65 g cm⁻³. This low density facilitates structural changes and phase transitions at high pressures. Finally, the high strength of the Si–O interatomic bond corresponds to the relatively high melting temperature of 1,723°C.

When crystalline silica is melted and then cooled, a disordered 3-dimensional network of silica tetrahedra (vitreous silica) is generally formed. Glass manufacturing in the USA is a 10 billion dollar per year industry. It directly benefits from studies of quartz as one of the main raw materials of commercial glasses is almost pure quartz sand, with other raw materials being primarily soda ash (Na_2CO_3) and calcite ($CaCO_3$) [1].

2.2 Quartz

Low (α) quartz allows little ionic substitution into its structure. High (β) quartz allows the charge-balanced substitution of framework silicon by aluminum, with a small cation (Li^+) occupying the interstices. In the more open cristobalite and tridymite structures, this charge-balanced substitution can be extensive, with many alkali and alkaline earth ions able to occupy interstitial sites. Such materials are called "stuffed silica derivatives," with eucryptite ($LiAlSiO_4$), nepheline ($Na_3K(AlSiO_4)_4$), carnegieite ($NaAlSiO_4$), and kalsilite ($KAlSiO_4$) being examples.

Similar to most other silica structures, quartz has a continuously connected network of $(SiO_4)^{4-}$ tetrahedra and an O/Si ratio equal to 2. This characteristic structure is also seen in cristobalite and tridymite. Interestingly, helices have been reported in quartz with two slightly different Si–O distances (0.1597 and 0.1617 nm) and an Si–O–Si angle of 144° [6]. Enantiomeric crystals of quartz are often obtained and separated mechanically. Each enantiomeric crystal of quartz is optically active.

According to Wyckoff [7], the crystalline forms of silica are the largest group of tetrahedral structures. Each of the three main polymorphs of silica formed at atmospheric pressure in nature (quartz, tridymite, and cristobalite) has a low and high temperature modification. The unit cell of low (α) quartz has three molecules and similar dimensions as the high (β) quartz structure. The difference between low and high forms of quartz arises from small shifts of atom positions. Table 1 lists structural data for both quartz structures.

The atomic arrangements in high and low quartz are very similar. In fact, when a single crystal of low quartz is carefully heated above 575°C, it is known to gradually and smoothly transform into a single crystal of high quartz, with a shift from a 3- to 6-fold symmetry [7]. The oxygen tetrahedron is almost regular (Si–O distance is 0.161 nm) for low quartz and with each oxygen having six adjacent oxygens (0.260–0.267 nm) and two silicon neighbors. Fourier analysis has provided accurate data for both structures [7]. The low and high forms of quartz are related by a displacive transformation with the former having the higher symmetry. Quartz, hexagonal in structure, is the lowest-temperature form of silica [5].

The structure of quartz has been extensively studied [7–10]. Table 2 summarizes structural data for low quartz obtained with the Accelrys *Catalysis 3.0.0* software. The continuous connection of oxygen tetrahedra is apparent from its structure illustrated in Figs. 4 and 5 [11,13].

Figure 5 shows that the linkage of tetrahedra in low quartz is, in fact, a double helix when viewed along the *a*-axis. This double helix structure was known long before the more celebrated structure of DNA [12,13].

Table 1 Comparison between low- and high-quartz structures [11] (after Wyckoff [7])

	Low temperature or α-quartz	High temperature or β-quartz
Bravais lattice	Hexagonal	Hexagonal
No. of ions	9 (3 Si^+, 6 O^{-2})	9 (3 Si^+, 6 O^{-2})
Temperature	<573°C[a]	573–867°C[a]
a_o	0.491304 nm	0.501 nm
c_o	0.540463 nm	0.547 nm
c/a	1.10[b]	1.09[b]
Space group	D_3^4 or D_3^6 ($P3_12$)[b]	D_6^4 or D_6^5 ($P6_22$)[b]
Si–O	0.161 nm	0.162 nm
O–O	0.260–0.267 nm	0.260 nm
Si–O–Si angle	144°[b]	155°[c]
Symmetry	Threefold	Sixfold
Molecules	3	3

Additional data as indicated from different references: [a]from [1], [b]from [8], and [c]from [6]

Table 2 Structure of low-quartz[a]

Bravais lattice	Unit cell dimensions[a] (*a*,*b*,*c*) in nm	Unit cell major angles[a] (α,β,γ)	Space group number	Symmetry number	No. of ions per unit cell
Hexagonal	0.49130, 0.49130, 0.54052	90.0, 90.0, 120.0	*P3₁21*	152	9

[a]From Accelrys software [11]; P = Primitive

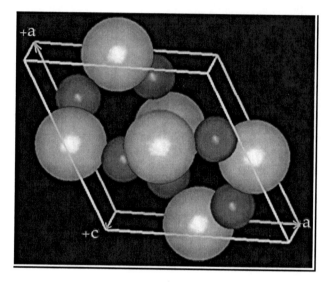

Fig. 4 Atomic arrangement in low-quartz (looking down the *c*-axis). (Small dark and larger light spheres represent oxygen and silicon ions respectively). The relative sizes of these ions correspond to the significant degree of covalency in the Si–O bond [11]

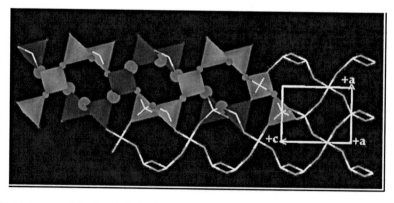

Fig. 5 Illustration of the double helix formed by SiO₄ tetrahedra in low-quartz (viewed down an *a*-axis) [11,13]

2.3 Cristobalite

Cristobalite, the highest-temperature polymorph of silica, was named after the place where it was discovered, the San Cristobal mountain in Mexico. Interestingly, silicate phases including cristobalite have also been found in cosmic dust collected by space vehicles [9]. The high cosmic and terrestrial abundance of silicas makes knowledge of their physical and chemical properties especially important in fields such as geology, chemistry, and physics. Cristobalite, like tridymite and keatite, is isostructural with ice polymorphs (i.e. cubic ice Ic).

Cristobalite has the Si atoms located as are the C atoms in diamond, with the O atoms midway between each pair of Si [13]. Like other crystalline polymorphs of silica, cristobalite is characterized by corner-shared SiO_4 tetrahedra. In addition, Liebau [9] noted that cristobalite, like quartz, exists in two forms having the same topology, with variations mainly in the Si–O–Si bond angles. Thermodynamic variables (such as pressure and temperature) and kinetic issues will determine which of these phases is formed. The interconversion of quartz and cristobalite on heating requires breaking and re-forming bonds, and consequently, the activation energy is high. However, the rates of conversion are strongly affected by the presence of impurities, or by the introduction of alkali metal oxides or other "mineralizers."

Cristobalite has been well-characterized since the late fifties [7,9]. The high-cristobalite structure is characterized by a continuously connected network of $(SiO_4)^{4-}$ tetrahedra and is summarized in Table 3. The atomic model of the high-cristobalite structure in Fig. 6 [11] was generated with Accelrys *Catalysis 3.0.0*. Also, the Si–O distances have been noted to range between 0.158 nm and 0.169 nm.

2.4 Vitreous Silica

Crystalline silicas contain ordered arrangements of anion tetrahedra, whereas glassy silica has a high degree of randomness. Comparisons of these networks indicate that both have the basic tetrahedral unit, the same O–Si–O bond angle (109.5°), an O/Si ratio of two, and full connectivity of tetrahedra. An equivalent short-range order has been found in both crystalline and glassy silica, as shown schematically in Fig. 7.

Three related structural parameters for characterizing the atomic-scale structure of vitreous silica are the Si–O–Si bond angle between adjacent tetrahedra, the rotational angle between adjacent tetrahedra, and the "rings" of oxygens, as illustrated in Fig. 7 [5]. Each of these parameters has a constant value or set of values in crystalline silica,

Table 3 Characteristics of high-cristobalite[a]

Bravais lattice	Unit cell dimensions[a] (a,b,c) in nm	Unit cell major angles[a] (α,β,γ) in degrees	Space group number	Symmetry number	No. of ions per cell
FCC	0.716, 0.716, 0.716	90.0, 90.0, 90.0	$Fd\bar{3}m$	227	24

[a]From Accelrys software [11]

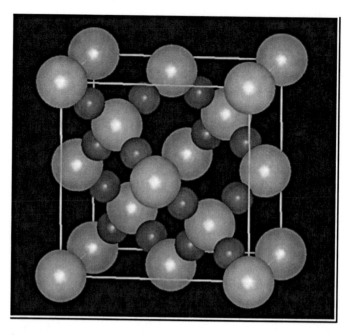

Fig. 6 Atomic arrangement in the high-cristobalite unit cell viewed down an *a*-axis. Small darker and large lighter spheres represent oxygen and silicon ions respectively. As in Fig. 4, the relative sizes of these ions correspond to the significant degree of covalency in the Si–O bond [11]

but varies over a wide range in vitreous silica. Table 4 summarizes these traits. The predominant Si–O–Si angle in quartz and cristobalite is 143.61° and 148°, respectively, and for tridymite it is 180° (one among a large group of angles). Vitreous silica, however, has a wide, continuous range of values between 120° and 180° (mean of less than 150°). The rotational angle between tetrahedra is either 0° or 60° for crystalline silica and is random in glass [2,5].

The common inorganic glasses used for windows and common glassware are silicates with significant amounts of oxides, other than SiO_2, present, such as Na_2O and CaO. Scientific glassware is generally a borosilicate containing B_2O_3, along with the soda and lime components. The boric oxide is a glass former, contributing to the oxide network polymerization, and glass modifiers (Na_2O and CaO) disrupt or depolymerize the network, reducing the melting and glass transition temperatures. Silica, as a chemical component in these glasses, is rather nonreactive to acids, H_2, Cl_2, and most metals at ordinary or slightly elevated temperatures, but it is attacked by fluorine, aqueous HF, and fused carbonates among others [14].

The general feature of vitreous silica as a continuously connected "random" network of SiO_4 tetrahedra was first defined by Zachariasen [15]. This nature of vitreous silica was verified by Warren et al. [16] within the limits of the X-ray diffraction techniques of that day. Several, subsequent studies have investigated the structure of vitreous silica and generally confirmed the open structure proposed by Zachariasen. Mozzi and Warren [17] substantially refined the X-ray work done by Warren et al. [16]

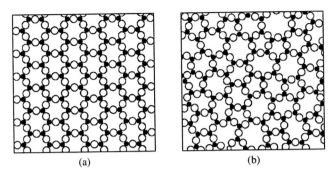

Fig. 7 Schematic 2-dimensional comparison of the structure of crystalline vs. noncrystalline silica [1]

Table 4 Some characteristics of crystalline and noncrystalline silica [2,5,11]

	SiO_2 glass	SiO_2 crystal
Number of nearest neighbors	Si: 4	Si: 4
	O: 2	O: 2
Bond Angles	109.5° (O–Si–O)	109.5° (O–Si–O)
	144° ± 15° rms[9]	180° (tridymite)[a]
		150.9°–143.61° (quartz)[b]
	(Si–O–Si)	Approx. 148° (cristobalite)[b]
Rotation angle between tetrahedra	Random	0° or 60°

[a]Only one among a large group of angles
[b]From [2]

identifying the average Si–O–Si bond angle at 144° and the overall distribution of that angle varying between 120° and 180°. Subsequent modeling studies largely confirmed the Mozzi and Warren results [18–20].

Until the 1950s, the Russian school of glass science favored a theory of the structure of vitreous silica based on the coincidence of the broad X-ray diffraction peaks for vitreous silica and the sharp peaks of cristobalite. The glass pattern was ascribed to line broadening due to the extremely small "particle size" [21] of such crystallites. However, for vitreous silica, this "microcrystallite" theory has largely been supplanted by the random network theory of Zachariasen. After more than seven decades, the Zachariasen model continues to be a very useful first-order description of vitreous silica. X-ray [17] and neutron [22] studies have generally supported this conclusion. On the other hand, silicate glasses with significant modifier content have provided evidence of subtle ordering effects analogous to crystalline silicates of similar composition. CaO–SiO_2 glass in comparison to wollastonite is an excellent example [23,24]. Figure 8 shows a computer-generated model of vitreous silica using well-established interatomic potentials for Si–O [25,26].

The short-range structure (i.e., length scale below 0.5 nm) of vitreous silica has been studied in terms of the structure factor and the radial distribution functions using neutron and X-ray diffraction experiments. Experimental radial distribution functions indicate that the separation distance between Si and O falls in the 0.159–0.162 nm range. The nearest neighbor distances O–O are 0.260–0.265 nm and the Si–Si distances are 0.305–0.322 nm [27,28].

High pressures can affect the properties of vitreous silica. For example, the nature of silica within the soil is a question of continuing inquiry in geology. Siliceous rocks that undergo meteorite impacts often form a detailed record of the high-pressure shocks on the surfaces. The response of vitreous silica to stress is also critical to technology, from tool making to the control of microstrains in modern nanolayered materials. High-pressure studies have unveiled a number of phenomena in silica glass, including the discovery of new phases, amorphization transitions, and unusual behavior under dynamic compression [29]. Thus, understanding the response of vitreous silica to high-pressure conditions has important implications for geology, planetary science, materials science, optics, and physics.

An indication of the effect of high pressure on the structure of vitreous silica is illustrated by the distortion of the ring size distribution. Shackelford and Masaryk [30] showed that the sizes of interstitial sites in vitreous silica follow a lognormal distribution. Similarly, the distribution of ring sizes in two-dimensional models of this material also follows the lognormal distribution [31]. Contemporary, rigorous three-dimensional simulations of vitreous silica (such as Fig. 8) clearly demonstrate this distribution. Figure 9 shows how this skewed distribution broadens significantly upon the application of high pressures. The average ring size in such structures at ambient conditions is six-membered (a loop of six connected silica tetrahedra), and the number of rings larger and smaller than six drops off sharply. Under high pressure, however, the number of six-membered rings is diminished and the relative numbers of larger and smaller rings (for example, eight- and four-membered rings) increase.

3   Key Properties of Quartz and Other Silicas

Quartz is abundant and hence inexpensive, relatively hard and chemically inert. Similar to other ceramics, high hardness is a useful property of quartz. Knoop hardness data for a number of ceramic materials including quartz are given in Table 5 [32]. The densities of a number of ceramic materials including quartz are given in Table 6 [32].

Extensive reviews have been reported on the mechanical behavior of vitreous silica [33]. The Young's modulus at 25°C is 73 GPa, the shear modulus is 31 GPa, and Poisson's ratio is reported as 0.17. Vitreous silica and silicates are notable solids because of their unique set of properties such as its ability to transmit visible light, ultraviolet and infrared radiation, good refractory and dielectric properties, chemical inertness, and low thermal expansion with resulting high thermal shock resistance. In the infrared region, water incorporated in the structure as hydroxyl (OH$^-$) has strong absorption bands at specific wavelengths. The Si–O vibration has two strong absorption bands that affect the transmission of silica. Transmission curves are typically compared

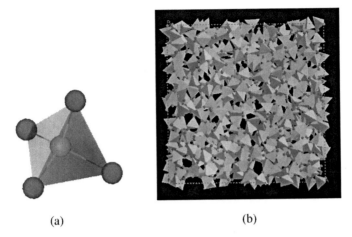

(a)　　　　　　　　　(b)

Fig. 8 The structure of vitreous silica is composed of a (a) basic building block, the $(SiO_4)^{-4}$ tetrahedron (corner spheres denote oxygen and the central sphere silicon), which forms (b) a 3D noncrystalline network of fully connected tetrahedra [26]

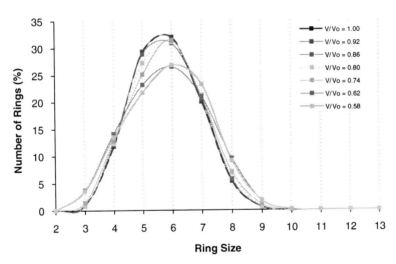

Fig. 9 The distribution of ring sizes in vitreous silica follows a lognormal distribution. The distribution broadens under increasingly high pressures

for several types of vitreous silica. Ultra-pure vitreous silica with elevated high transparency is required in telecommunication fiber optics.

Different sources of radiation can affect the physical and optical properties of vitreous silica. For instance, a dose of 1×10^{20} neutrons cm^{-2} has been reported to increase the density of vitreous silica by about 3% (to 2.26 g cm^{-3}) [27]. Similar increases in density are reported in quartz, tridymite, and cristobalite after comparable irradiation levels. On the other hand, ionizing radiation such as X-rays, γ-rays, electrons, or protons carry

Table 5 Knoop hardness for quartz and some common ceramic materials [32]

Material	Knoop hardness (100 g load) (in kg mm^{-2})
Boron carbide (B$_4$C)	2800
Silicon carbide (SiC)	2,500–2,550
Tungsten carbide (WC)	1,870–1,880
Aluminum oxide (Al$_2$O$_3$)	2,000–2,050
Zirconium oxide (ZrO$_2$)	1,200
Quartz (SiO$_2$)	
Parallel to optical axis	710
Perpendicular to optical axis	790
SiO$_2$ glass	500–679

Table 6 Densities for quartz and some common ceramic materials [32]

Material	Density (g cm^{-3})
Boron carbide (B$_4$C)	2.51
Silicon carbide (SiC)	
Hex.	3.217
Cub.	3.210
Tungsten carbide (WC)	15.8
Aluminum oxide (Al$_2$O$_3$)	3.97–3.986
Zirconium oxide (ZrO$_2$)	5.56
Quartz (SiO$_2$)	2.65
SiO$_2$ glass	2.201–2.211

enough energy to produce absorption centers, known as color or defect centers, in vitreous silica. A characteristic band at 215 nm is produced by long exposures to X-radiation [27]. This band is also reported in irradiated α-quartz and is often associated with the existence of E′ centers, a type of defect assumed to be a pyramidal SiO$_3$ unit having an unpaired electron in the Si sp^3 orbital. Various types of defect centers in silica glass can be classified as either intrinsic (melt-quench) or extrinsic (radiation-induced).

4 Processing Quartz and Other Silicas

Silica, the main component of silicates, is widely used as mentioned earlier. In its crystalline and noncrystalline polymorphs, silica is used industrially as a raw material for glasses, ceramics, foundry molds, in the production of silicon, and more recently in technical applications such as quartz oscillators and optical waveguides for long-distance telecommunications. Of the crystalline forms, only α-quartz is commonly used as sand or as natural and synthetic single crystals. Cristobalite is often utilized as the synthetic phase in glass-ceramics.

Beyond the abundant natural sources of quartz and other silicas, techniques for synthetic production of these materials have provided a significantly wider range of applications [27,34]. Large, high-quality crystals of quartz can be grown by the well-established technique of hydrothermal growth in an autoclave filled with a solution of sodium carbonate at elevated temperature and pressure. Quartz particles are fed into the bottom of the growing chamber, while seed crystals are fed into the top in a metal frame. A temperature gradient establishes a greater solubility at the higher-temperature bottom of the chamber, leading to a continuous transfer of material upward to the growing single crystals. Uniform quality crystals are routinely produced with well-controlled shapes and sizes. Specific seed crystal orientations are used to produce desired products such as particular oscillator configurations.

Vitreous silica has a unique set of properties for applications where optical transmission, chemical inertness, and thermal stability are crucial. The abundance of vitreous silica in nature is widespread in biogenic sources such as sponges and diatoms, in crystalline opals, and as glass cycled by organisms through the environment (e.g., silicification of plant tissues for structural integrity and protection from insects [35]). This important glass can also be readily found in abiogenic sources such as volcanic glasses, resulting from extensive quenched magmas, tektites (spherical or teardrop-shaped silicate glass bodies linked with impact craters), and lechatelierite (pure silica glass), resulting from lightening strikes of unconsolidated sand or soil that form fulgurites. Glassy silica is also formed by a combination of temperature and pressure resulting from meteoritic impact [27].

Vitreous silica is high purity SiO_2 glass that can withstand service temperatures above 1,000°C. As a metastable phase of silica, vitreous silica can be readily obtained in nature and synthetically. Silica glass can be produced in a pure and stable form, displaying useful properties, but is rigid and difficult to shape even at 2,000°C. Hence, it is not accessible to mass production plastic-forming methods. However, techniques have been developed to produce vitreous silica in various shapes and sizes [36–39].

First, quartz crystals can be melted to produce silica glass by either the Osram process or the Heraeus method [27]. In the Osram technique, fragmented quartz is fed to a tubular furnace and melted in a crucible protected by an inert gas, where tubing is drawn from the bottom of the crucible. In the Heraeus method, quartz crystals are fed in an oxy-hydrogen flame through a rotating fused quartz tube and withdrawn slowly from the burner as clear fused (vitreous) silica accumulates. The quartz crystals are generally washed in hydrofluoric acid and distilled water to remove surface impurities, followed by drying and heating to ~800°C, before being immersed in distilled water. Purity of the natural sand is very important in glass and ceramic materials, and transition metal oxides should not exceed 200 ppm.

In vapor phase hydrolysis [37,38], synthetic vitreous silica is prepared from silicon tetrachloride by oxidation or hydrolysis in a methane–oxygen flame. The resulting soot is sintered to form silica glass. Water, formed from the oxidation of methane, subsequently combines with the chloride, leading to the production of hydrochloric acid and oxygen. Subsequent work on these materials can lead to a variety of useful products, including telescope mirror blanks, lamp tubing, crucibles, and optical fibers (the largest commercial use for vitreous silica in telecommunications).

Finally, vitreous silica can be manufactured by the sol–gel technique developed by Zarzycki [39]. Gels are formed by the destabilization of colloidal sols or by the hydrolysis of metal organic compounds. This latter routine is the most common technique that yields a silica–alcohol–water gel. Subsequently, the gel is dried and fused to produce silica glass. The manufacture of 3D articles by this method is limited due to the difficulty in drying porous gels without large shrinkage and cracking and the associated high costs of the raw materials. Silica gel can also be used as a drying agent and as supports for chromatography and catalysis [6].

5 Future Trends

Various silicas, including quartz, are especially interesting in that they represent a family of materials that are familiar, while also providing state-of-the-art applications. As an example of the commonplace, the largest part of the industrial sand and gravel production in the United States (39% in 2004 corresponding to more than ten million tons) is glassmaking sand [40]. This important raw material is the relatively high-purity quartz with only small amounts of alumina and iron oxide impurities permitted. Health and safety regulations are expected to cause future sand and gravel operations to be relocated to areas more remote from high-population centers.

As noted earlier in this chapter, vitreous silica is used increasingly in a number of advanced applications such as fiber optics, laser systems, and waveguides. In addition, vitreous silica continues to be an excellent model system for the study of the structure of noncrystalline solids. One can expect that the continuing refinement of our understanding of this structure will be aided by the availability of a new generation of diffraction systems at the Spallation Neutron Source (SNS) at the Oak Ridge National Laboratory and the GLAD diffractometer at the Argon National Laboratory. Much of the focus of these structural studies as well as future technological applications will be the "medium-range" nano-scale that exists between the short-range order of the silica tetrahedron and the long-range randomness of vitreous silica. Computer simulations have played a key role in predicting the nature of such length scales in this important glassy material [11,26,41]. Further improvements of interatomic potentials and computing power will certainly expand our understanding of this material and perhaps one-day allow the design of ceramics and glasses with specific, desirable properties not currently available.

Acknowledgments One of the authors (LPD) performed much of her work with the support of a Student Employee Graduate Research Fellowship (SEGRF) from the Lawrence Livermore National Laboratory.

References

1. J.F. Shackelford, *Introduction to Materials Science for Engineers*, 6th edn., Prentice-Hall, Upper Saddle River, NJ, 2005.
2. P.J. Heaney, C.T. Prewitt, and G.V. Gibbs (eds.), *Silica: Physical Behavior, Geochemistry and Materials Applications, Reviews in Mineralogy*, Vol. 29, Mineralogical Society of America, Washington, DC, 1994.
3. J.W. Chan, T.R. Huser, S.H. Risbud, and D.M. Krol, Modification of the fused silica network associated with waveguide fabrication using femtosecond laser pulses, *Appl. Phys. A*, **76**, 367–372 (2003).

4. C. Klein and C.S. Hurlbut, Jr., *Manual of Mineralogy,* 21st edn., John Wiley, NY, 1993, p. 527.
5. Y.M. Chiang, D.P. Birnie, III, and W.D. Kingery, *Physical Ceramics: Principles of Ceramic Science and Engineering,* John Wiley, New York, 1997.
6. R.B. King, *Inorganic Chemistry of Main Group Elements,* VCH, Berlin, 1995.
7. R.W.G. Wyckoff, *Crystal Structures,* Interscience, New York, 1957 and 1959.
8. E. Parthe, *Crystal Chemistry of Tetrahedral Structures,* Gordon and Breach, New York, 1964.
9. F. Liebau, *Structural Chemistry of Silicates,* Springer, Berlin, 1985.
10. R.B. Sosman, *The Phases of Silica,* Rutgers University Press, Piscataway, NJ, 1965.
11. L.P. Davila, Computer modeling studies of the interstitial structure of selected silica polymorphs, M.S. Thesis, University of California, Davis, 1998.
12. P.J. Heaney, Structure and chemistry of the low-pressure silica polymorphs, in *Silica: Physical Behavior, Geochemistry and Materials Applications, Reviews in Mineralogy,* Vol. 29, by P.J. Heaney, C.T. Prewitt, and G.V. Gibbs (eds.), Mineralogical Society of America, Washington, DC, 1994, pp. 1–40.
13. L.P. Davila, S.H. Risbud, and J.F. Shackelford, Quantifying the interstitital structure of non-crystalline Solids, *Recent Res. Devel. Non-Cryst. Solids,* **1,** 73–84 (2001).
14. F.A. Cotton and G. Wilkinson, *Adv. Inorganic Chem.,* 4th (edn.), John Wiley, New York, 1980.
15. W.H. Zachariasen, The atomic arrangement in glass, *J. Am. Chem. Soc.* **54,** 3841–3851 (1932).
16. B.E. Warren et al., Fourier analysis of X-ray patterns of vitreous SiO_2 and B_2O_3, *J. Am. Chem. Soc.* **19,** 202–206 (1936).
17. R.L. Mozzi and B.E. Warren, The structure of vitreous silica, *J. Appl. Cryst.* **2,** 164–172 (1969).
18. R.J. Bell and P. Dean, The structure of vitreous silica: validity of the random network theory, *Phil. Mag.* **25,** 1381–1398 (1972).
19. J.R.G. da Silva, *et al.,* A refinement of the structure of vitreous silica, *Phil. Mag.* **31,** 713–717 (1975).
20. P.G. Coombs et al., The nature of the Si–O–Si bond angle distribution in vitreous silica, *Phil. Mag.* **51,** L39–L42 (1985).
21. J.T. Randall, H.P. Rooksby, and B.S. Cooper, X-ray diffraction and the structure of vitreous solids – I, *Z. Krist.* **75,** 196–214 (1930).
22. A.C. Wright, R.A. Hulme, D.I. Grimley, R.N. Sinclair, S.W. Martin, D.L. Price, and F.L. Galeener, The structure of some simple amorphous network solids revisited, *J. Non-Cryst. Sol.* **129,** 213–232 (1991).
23. P.H. Gaskell, M.C. Eckersley, A.C. Barnes, and P. Chieux, Medium-range order in the cation distribution of a calcium silicate glass, *Nature* **350,** 675–677 (1991).
24. M.C. Abramo, C. Caccamo, and G. Pizzimenti, Structural properties and medium-range order in calcium-metasilicate ($CaSiO_3$) glass: a molecular dynamics study, *J. Chem. Phys.* **96,** 9083–9091 (1992).
25. B.P. Fueston and S.H. Garofalini, Empirical three-body potential for vitreous silica, *J. Chem. Phys.* **89,** 5818–5824 (1988).
26. L.P. Davila, Atomistic-scale simulations of vitreous silica under high pressure: structure and properties, Ph.D. Dissertation, University of California, Davis, 2005.
27. G.H. Beall, Industrial applications of silica, in *Silica: Physical Behavior, Geochemistry and Materials Applications, Reviews in Mineralogy,* Vol. 29, P.J.Heaney, C.T. Prewitt, and G.V. Gibbs (eds.), Mineralogical Society of America, Washington, DC, 1994, pp. 469–505.
28. C.J. Simmons and O.H. El-Bayoumi, *Experimental Techniques of Glass Science,* The American Ceramic Society, Westerville, OH, 1993.
29. R.J. Hemley, C.T. Prewitt, and K.J. Kingma, High-pressure behavior of silica, in *Silica: Physical Behavior, Geochemistry and Materials Applications, Reviews in Mineralogy,* Vol. 29, P.J.Heaney, C.T. Prewitt, and G.V. Gibbs (eds.), Mineralogical Society of America, Washington, DC, 1994, pp. 41–81.
30. J.F. Shackelford and J.S. Masaryk, The interstitial structure of vitreous silica, *J. Non-Cryst. Solids* **30,** 127–139 (1978).
31. J.F. Shackelford and B.D. Brown, Triangle rafts – extended Zachariasen schematics for structure modeling, *J. Non-Cryst. Solids,* **49,** 19–28 (1982).
32. J.F. Shackelford and W. Alexander, *Materials Science and Engineering Handbook,* 3rd edn., CRC Press, Boca Raton, FL, 2001.

33. R. Bruckner, Mechanical properties of glasses, in *Glasses and Amorphous Materials*, R.W. Cahn, P. Haasen, and E.J. Kramer (eds.), VCH, New York, 1991, pp. 665–713.

34. W.D. Kingery, *Introduction to Ceramics*, John Wiley, New York, 1960.

35. C.H. Chen and J. Lewin, Silicon as a nutrient element for *Equisetum Arvense*, *Can. J. Bot.* **47**, 125–131 (1969).

36. P. Danielson, Vitreous silica, in *Encyclopedia of Chemical Technology*, Vol. 20, John Wiley, New York, 1982, pp. 782–817.

37. F.P. Kapron, D.B. Keck, and R.D. Maurer, Radiation loss in glass optical waveguide, *Appl. Phys. Lett.* **17**, 423–425 (1970).

38. D.B. Keck, R.D. Maurer, and P.C. Schultz, On the lower limit of attenuation in glass optical waveguides, *Appl. Phys. Lett.* **22**, 307–309 (1973).

39. J. Zarzycki, Glasses and the vitreous state, in *Cambridge Solid State Science Series*, R.W. Cahn, E.A. Davies, and I.M. Ward (eds.), Cambridge University Press, Cambridge, UK, 1991, pp. 194–196.

40. T.P. Dolley, Materials review: sand and gravel (industrial), *Am. Ceram. Soc. Bull.*, **84**, 24 (2005).

41. L.P. Davila et al. Transformations in the medium-range order of fused silica under high pressure, *Phys. Rev. Lett.* **91**, 205501-1–205501-4 (2003).

Chapter 6
Refractory Oxides

Jeffrey D. Smith and William G. Fahrenholtz

Abstract Refractory oxides encompass a broad range of unary, binary, and ternary ceramic compounds that can be used in structural, insulating, and other applications. The chemical bonds that provide cohesive energy to the crystalline solids also influence properties such as thermal expansion coefficient, thermal conductivity, elastic modulus, and heat capacity. This chapter provides a historical perspective on the use of refractory oxide materials, reviews applications for refractory oxides, overviews fundamental structure–property relations, describes typical processing routes, and summarizes the properties of these materials.

1 Introduction

The term *refractory* refers to materials that are resistant to the effects of heat. Refractory oxides, therefore, are ceramic materials that can be used at elevated temperatures. These nondescript restrictions allow nearly any oxide to be classified as refractory. For this article, refractory oxides will refer, somewhat arbitrarily, to common crystalline compounds with melting temperatures of at least 1,800°C. These compounds can contain one or more metal or metalloid cations bonded to oxygen. As an introduction to the topic, this section provides a brief historic overview of materials commonly used in the refractories industry, including some lower melting temperature materials. The section also reviews some current trends in the industries that produce and use refractory oxides. The other sections of this chapter focus on phase-pure oxide ceramics that can be used at elevated temperatures.

Historically, most of the oxides that were used in refractory applications were traditional ceramics prepared from clays or other readily available mineral-based raw materials. The major categories of traditional refractories are fire clays, high aluminas, and silica [1]. The choice of material for traditional refractory applications, as with advanced material applications, was and is based on balancing cost and performance/lifetime. The ultimate use temperatures and applications for some common refractories are summarized in Table 1 [2, 3]. The production, properties, and uses of some of these materials are discussed in more detail in the other chapters

J.F. Shackelford and R.H. Doremus (eds.), *Ceramic and Glass Materials: Structure, Properties and Processing.*

Table 1 Compositions, ultimate use temperatures, and applications for some common traditional refractory materials

Class	Material	Phases	Use Temp (°C)	Applications
Fire clay	Low heat duty	Mullite, glass, quartz	Up to 1500	Kiln linings
	High heat duty	Mullite, glass		Crucibles
High alumina	Kyanite	$\alpha\text{-}Al_2O_3$, mullite, glass	Up to 1800	Metal handling
				Lab ware
Silica	Silica	Tridymite,	1650	Glass tanks
		cristobalite		crowns

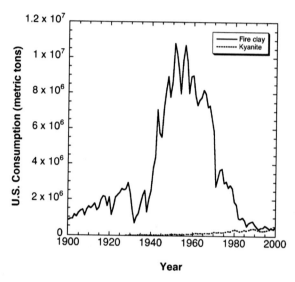

Fig. 1 Historic production numbers for fireclay and high alumina (labeled kyanite) brick

of this volume. A brief overview of fireclays, high aluminas, and silica is provided here followed by a description of the evolution of the refractories industry.

Although no strict geologic definition exists [4], fireclays can be defined as clay minerals that have pyrometric cone equivalent (PCE) values of 19 or greater following ASTM specification C24–01 (Standard Test Method for Pyrometric Cone Equivalent (PCE) of Fireclay and High Alumina Refractory Materials) [5]. Most refractory products are fabricated from what are considered high heat duty fire clays, which have a PCE value of 27 or higher (~1600°C). Fireclays have Al_2O_3 contents that range from 20 to 45 wt%, with silica being the other major constituent [6]. Because of their ease of fabrication, resistance to chemical attack, and low cost, fireclay bricks are still widely used as refractory materials. Applications for fireclay refractory brick include insulation behind hot-face materials, furnace linings, and specialty applications such as laboratory crucibles and setters. Historic consumption of fireclay was significantly greater when fireclay refractory brick demand from the U.S. steel industry was at its peak of ~10,000,000 metric tons in the early 1950s (Fig. 1). The decline in demand from the steel industry was due to changes that included higher use temperatures and a shift to basic practices to improve steel cleanliness. The changing process requirements spurred the development of advanced refractory ceramics such as high alumina castables and basic brick, both of which are prepared from highly beneficiated oxides rather than unrefined minerals. In the past quarter century, fireclay refractories have evolved from a state-of-the-art engineered material to a commodity item that often originates from countries having low labor costs.

Most high alumina refractories are clay-based ceramics to which an alumina-rich mineral is added to chemically react with a majority of the silica present to promote mullite formation [7]. High alumina refractories contain a minimum of 60 wt% Al_2O_3, although the Al_2O_3 content can be >99% for specialty products. High alumina refractories can be produced from fire clays used in combination with alumina-rich minerals such as diaspore or bauxite [8]. Reduction of the amount of free silica (consumed in the formation of mullite) results in increased use temperature for high alumina refractories compared with fire clay refractories, up to 1800°C for some materials. The greater mullite content of high aluminas gives them improved creep resistance and better corrosion behavior. High alumina refractories were developed for steel industry applications that were beyond the performance limits of fireclay refractories. High alumina bricks continue to find use in a wide range of industrial applications including aluminum melting and incineration. Today, use of high alumina materials is approximately equivalent to fireclays (Fig. 1).

Silica refractories can be crystalline or amorphous (fused). Most silica refractories are produced from silica-rich minerals such as quartz and flint and have SiO_2 contents of 98 wt% or higher. For crystalline refractories, a mineralizer-like CaO is added to promote crystallization to cristobalite and/or tridymite thereby eliminating the displacive-phase transformation associated with the α to β quartz transition at 573°C. Displacive transformations are typically associated with substantial volume changes that can be quite destructive. Because of the relatively low theoretical density of silica (~2.3 g cm^{-3} for cristobalite and tridymite), silica bricks are often used to construct arched furnace crowns [8]. Unlike most ceramic materials, silica bricks are resistant to creep at elevated temperature allowing them to be used for extended durations at temperatures approaching the melting temperature. Thus, even though silica melts below the 1800°C limit considered in this article, it has been included because of its high use temperature. The recent trend in the glass industry to convert to oxy-fuel firing has decreased the usage of silica brick because higher temperatures and water vapor concentration in oxy-fuel fired glass hearths promotes alkali-induced corrosion of silica.

In the middle part of the twentieth century, the ceramics industry began a general shift from traditional ceramics toward more advanced (highly engineered) materials. Traditional ceramics are derived from minerals and can have significant variations in composition and performance depending upon the source of the raw material. Traditional ceramics also tend to contain significant amounts of glassy phases or impurities. In contrast, advanced ceramics are usually phase pure oxides that are derived from high-purity industrial chemicals. Advanced ceramics can be single phase or multiphase, but they are essentially phase pure meaning that they contain no significant (0.5 wt% or less) glassy phase or impurities. The cost of advanced ceramics compared with traditional materials created the need for application-specific compositions. Thus, advanced materials are implemented specifically where they are needed to optimize system performance. The selection of advanced materials is still driven by the performance-cost balance. Understanding materials performance and selecting the proper material for a particular application requires knowledge of material properties such as those discussed later in this chapter.

Even though many refractory oxides are engineered to optimize performance in a single application, any number of ceramics can be selected for a particular application. Examples of some of the oxides that can be used at high temperatures, along with their melting temperatures, are listed in Tables 2–5 for oxides containing one, two, or more cations [9–11]. It should be noted that consensus on the melting temperature of specific oxides is tenuous, so values should be considered as approximations; this is especially true in the case of oxides having melting temperatures well above 2000°C. These lists are not intended to be comprehensive (although Tables 2 and 5 contain all of the unary and ternary refractory oxides that the authors could identify), but the lists are long enough to emphasize that a large number of candidates exist for any application. Tables 3 and 4 are samplings from the hundreds of two component refractory oxides that are available.

From the larger list of binary refractory oxides, aluminate compounds are listed in Table 3 to emphasize that a family of materials that contain one compound with a high melting temperature will tend to form other compounds with high melting temperatures. Within the aluminate family, a number of compounds are formed that might not

Table 2 Melting temperatures of refractory oxides containing a single cation

Oxide	T_m (°C)
Al_2O_3	2020
BaO	1925
BeO	2570
CaO	2600
CeO_2	2600
Cr_2O_3	2400
CuO	1800
Eu_2O_3	2240
Gd_2O_3	2350
HfO_2	2780
In_2O_3	1910
La_2O_3	2315
MgO	2800
MnO	1815
NbO_2	1915
Nd_2O_3	2275
NiO	1960
Sc_2O_3	2450
Sm_2O_3	2310
SrO	2450
Ta_2O_5	1875
ThO_2	3250
TiO_2	1850
Ti_2O_3	2130
UO_2	2750
U_2O_3	1975
Y_2O_3	2400
Yb_2O_3	2375
ZnO	1975
ZrO_2	2700

Table 3 Melting temperatures of selected aluminates

Oxide	T_m (°C)
$BaO·Al_2O_3$	2,000
$BeO·Al_2O_3$	1,910
$CaO·6Al_2O_3$	1,850
$CeO·Al_2O_3$	2,070
$CoO·Al_2O_3$	1,955
$FeO·Al_2O_3$	1,820
$K_2O·Al_2O_3$	2,260
$La_2O_3·Al_2O_3$	2,100
$Li_2O·5Al_2O_3$	1,975
$MgO·Al_2O_3$	2,135
$Na_2O·11Al_2O_3$	2,000
$NiO·Al_2O_3$	2,020
$SrO·Al_2O_3$	1,960
$Y_2O_3·Al_2O_3$	1,940
$ZnO·Al_2O_3$	1,950

Table 4 Melting temperatures of barium-containing binary refractory oxides

Oxide	T_m (°C)
$BaO·Al_2O_3$	2,000
$3BaO·2Dy_2O_3$	2,050
$2BaO·GeO_2$	1,835
$6BaO·Nb_2O_5$	1,925
$BaO·Sc_2O_3$	2,100
$2BaO·SiO_2$	1,820
$BaO·ThO_2$	2,300
$2BaO·TiO_2$	1,860
$BaO·UO_2$	2,450
$3BaO·2Y_2O_3$	2,160
$BaO·ZrO_2$	2,700

Table 5 Melting temperatures of ternary refractory oxides

Oxide	T_m (°C)
$2CaO·Y_2O_3·Al_2O_3$	1,810
$Na_2O·9Y_2O_3·12SiO_2$	1,850
$2CaO·Gd_2O_3·Al_2O_3$	1,830
$3Ga_2O_3·2Sc_2O_3·3Al_2O_3$	1,850
$ZnO·ZrO_2·SiO_2$	2,080

normally be expected to be refractory such as those containing potassium oxide, sodium oxide, and even lithium oxide. Individually, oxides such as Li_2O, Na_2O, and K_2O would never be considered refractory, but combined with aluminum oxide they form refractory compounds.

The binary oxides listed in Table 4 were intended as a compilation that is similar to what was presented in Table 3. However, in this case barium oxide was chosen as one component of the binary system. Barium oxide is refractory (Table 2) and forms binary refractory compounds in a number of different families, include aluminates, silicates, titanates, and zirconates. Although not absolute, it is common for an oxide that is refractory in one family of oxides to be refractory in others as well.

Looking toward the future, it is likely that the current trends in production and use of high temperature materials will continue. The users of high temperature structural materials continually push for higher use temperatures and improved component lifetime. As use temperatures increase, it is likely that alternate materials that are now considered exotic will have to be developed; this development will be application specific and will occur at a rate that often lags the rate of process development. Consider the thoughts of a steel mill operator from the early 1900s if he had been told that in 50 years his plant would use basic refractories costing orders of magnitude more than fireclay brick. Other developments that are likely in the refractory materials field are the increased use of multiphase materials and coatings. Both technologies offer the promise of unique combinations of physical and mechanical properties that are not available in single-phase materials. For example, a multiphase engineered material could be constructed to have the wear resistance of a hard ceramic with the thermal conductivity and thermal shock resistance of a metal. The possible combinations of properties are nearly endless, but development of these materials requires knowledge of interactions at bimaterial interfaces, tailoring of thermal expansion coefficients, and development of cost-effective processing routes.

The purpose of this chapter is to describe the properties and applications for refractory oxides. The sections that follow describe applications, review fundamental chemical and physical aspects, introduce processing methods, list important physical properties, and discuss materials selection criteria for refractory oxides. The organization of this chapter reflects that the performance of ceramic materials depend on interrelationships among structure, processing, and inherent properties.

2 Applications

Oxides are used by refractory and structural ceramics manufacturers to produce materials that are used in a wide variety of industries. Even with the reduced production of steel in the US, the industry continues to be the largest (in terms of tonnage) consumer of refractory products. The high temperatures required for domestic steel production coupled with increasingly stringent performance demands and ever-present cost concerns continue to drive development of new products. Annually, the steel industry consumes about one-half of the World's refractory materials. The next two largest consumers of refractories, the aluminum and the glass industries, only account for about 20% of the refractory materials produced.

Remaining production and usage is distributed over a host of industries, many of which are not commonly known. Others include nonferrous metal producers (copper, lead, zinc, etc.), the cement industry, petroleum and hydrocarbon refineries, chemical producers, pulp and paper, food production-related industries; anything involving heat and/or hot products. Although only a minor consumer, NASA utilizes refractory tiles to protect astronauts from the harsh conditions that exist on operation of the space shuttle and a refractory brick pad to manage the heat load associated with launch.

The specific application defines the type of refractory material that can be utilized not only by property requirements but also by cost requirements. Each of the industries mentioned balances refractory performance with refractory cost. At times higher quality oxide refractories are abandoned in favor of less costly, but also less affective alternatives. As these industries continue to evolve to higher and higher production temperatures, acceptable lower cost alternatives will become increasingly less available.

3 Fundamental Explanations

The properties of metal oxide compounds depend on the individual atoms present, the nature of the bonding between the atoms, and the crystalline structure of the resulting compound. Materials engineers are concerned with the physical manifestations of bonding and crystal structure, meaning macroscopic properties such as elastic modulus and coefficient of thermal expansion, rather than the nature of the interactions among atoms. However, the ability to tailor material behavior and to design compositions and microstructures for specific applications requires an understanding of the fundamental physical and chemical principles that control bonding and crystal structure. To address these points, this section provides a brief review of atomic structure and bonding, crystal structure, and the resulting macroscopic behavior as they pertain to oxide ceramics.

3.1 Atomic Structure and Bonding

On the atomic level, the arrangement of electrons surrounding a nucleus determines how a particular atom will interact with other atoms [12]. The modern understanding of electronic structure is built on the concept of the Bohr atom extended to atoms with many electrons using the principles of quantum mechanics [13]. Each electron that surrounds a particular atom has a set of four quantum numbers that designates its shell (principal quantum number $n = 1, 2, 3$, etc.), its orbital (l = integer with values ranging from 0 to $n - 1$ representing the s, p, d, and f orbitals), its orientation (m_l = integer with values from $-l$ to $+l$), and its spin ($m_s = +1/2$ or $-1/2$). By the Pauli exclusion principal, each electron surrounding an atom has a unique set of four quantum numbers [14]. Standard versions of the periodic table are arranged in rows according to the electronic shell that is filled as the atomic number increases [15]. For example, atoms in the first row of the periodic table (H and He) have electrons in the first shell ($n = 1$). The increasing number of species in the lower rows of the periodic table results from the increased number of orbitals available for occupancy as n, the principal quantum number, increases. The columns represent groups of atoms with the same outer shell configuration. For example, the atoms in column IA (H, Li, Na, K, etc.) have one electron in the s orbital of the outermost shell.

The outermost electron shell surrounding an atom is referred to as its valence shell and it is the valence shell electrons that participate in chemical bonding [12]. Most often, it is the s and p orbital electrons (orbital quantum numbers 0 and 1) that affect the strength and directionality of chemical bonds [13]. When bonding, atoms minimize their energy by gaining, losing, or sharing electrons in an attempt to attain the electronic structure of the inert gas with the closest atomic number. When atoms gain or lose electrons they become ions. Ions with opposite charges form what are termed ionic bonds. If electrons are shared, directional covalent bonds are formed. Conversely, ionic bonding is nondirectional and the resulting solids tend to have high (6, 8, or 12) cation coordination numbers. For example, CsCl is an ionic compound composed of Cs^+ and Cl^- ions. Each Cs^+ cation is surrounded by eight Cl^- anions. Covalent bonds are directional based on the shape of the electron orbitals or the type of hybrid orbital that is formed to facilitate electron sharing [13]. Covalent compounds tend to have lower cation coordination numbers (3 or 4) compared with ionic compounds. An example of a covalent compound is SiC, in which each Si atom is bound to four C atoms and the angle between each bond is ~109°, and the angle of separation for sp^3 hybrid orbitals that is also known as the tetrahedral angle. In real oxide compounds, the bonds have both ionic and covalent characteristics. These bonds are referred to as iono-covalent or polar covalent [13, 16]. The degree of ionic character can be estimated using a variety of means including Pauling's electronegativity scale, Sanderson's model, or Mooser-Pearson plots [13]. Oxides are not generally close-packed like compounds that are predominantly ionic, but are not as open as highly covalent compounds.

Regardless of the type of chemical bond that forms, the net force between two chemically bound atoms results from electrostatic attraction [16]. The attractive component, E_{attr}, of the total bond energy between two atoms is a function of the distance between them, r. The normal form of the attractive force, based on Coulomb's law, for ionic crystals is

$$E_{attr} = \frac{z_1 z_2 e^2}{4\pi\varepsilon_0 r},$$

(1)

where z_1 and z_2 are the valences of the two atoms, e is the charge on an electron (1.602×10^{-19}C), and ε_0 is the permittivity of free space (8.854×10^{-12} C^2 N^{-1} m^{-2}).

The attractive energy acts over long ranges and can take slightly different forms for covalent bonding [13]. Without a repulsive force to balance the attractive force, all of the atoms in the universe would eventually be drawn into a single mass of infinite density. Fortunately, as atoms approach each other, a short-range electrostatic repulsion builds due to the overlap of the charge distributions from the two atoms [15]. Most often, the repulsive energy is expressed as the Born repulsion:

$$E_{rep} = \frac{B}{r^n},$$

(2)

where B is an empirical constant and n is the Born exponent, also an empirical constant, usually between 6 and 12.

The net energy between two atoms is the sum of the attractive and repulsive energies [15]. The equilibrium atomic separation, r_0, occurs at the point where the net energy shows a maximum in attraction. The value of r_0 can be calculated by taking the first derivative of the net energy, setting it equal to zero, and solving for r. A representative plot of the attractive, repulsive, and net energies is shown in Fig. 2. The magnitude of the maximum in the attractive energy determines the bond strength and, therefore, the lattice energy, of a crystal. Considering compounds with the same structure, differences in lattice energy affect macroscopic properties [13]. An example comparing the lattice energies, melting temperatures, and thermal expansion coefficients of alkaline

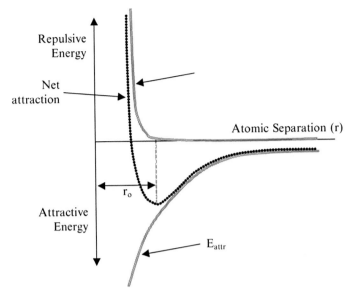

Fig. 2 Attractive, repulsive, and net interatomic energy as a function of interatomic separation

Table 6 Lattice energy, melting temperature, thermal expansion, and modulus for alkaline earth oxides with the rock salt structure

	Lattice energy (kJ mol^{-1})	Melting temperature (°C)	Coefficient of thermal expansion (ppm°C^{-1})
MgO	3,938	2,852	10.5
CaO	3,566	2,614	11.2
SrO	3,369	2,430	14.2

earth oxides that have the rock salt structure (MgO, CaO, and SrO) is outlined in Table 6 [13, 17, 18]. As observed by the trend in the data, melting temperature tends to increase and thermal expansion coefficient tends to decrease as the cohesive force, expressed as the lattice energy in this example, increases.

3.2 Crystal Structure

On the nanometer level, crystal structures are symmetric arrangements of molecules (bound atoms) in three-dimensional space [19]. Driven purely by energy minimization, countless manifestations of symmetry are found in nature ranging from the arrangement of atoms in unit cells and water molecules in snowflakes to the facets of crystals such as quartz and diamond [20]. For a crystal constructed of identical molecules, the positions of all of the molecules in the structure can be predicted using four basic symmetry elements: (1) centers of symmetry; (2) two, three, four, or sixfold rotational axes; (3) mirror or reflection planes; or (4) combinations of a symmetry centers and rotational axes [21]. Combined with the constraint that space must be filled by the resulting structural units, the symmetry elements can be used to construct structures that make up the seven basic crystal systems (cubic, hexagonal, rhombahedral, tetragonal, orthorhombic, monoclinic, and triclinic). Within the crystal systems, increasingly finer divisions of symmetry can be defined using Bravais lattices, crystal classes, or space groups (Table 7) [22]. A detailed description of how the symmetry elements relate to this hierarchy can be found in many texts on crystallography [19, 23], X-ray diffraction [21, 22], or mineralogy [20, 24]. As an aside, the convention is to name crystal structures after the mineral for which the positions of the atoms were first confirmed [16]. Thus, compounds showing face-centered cubic symmetry and belonging to the Fm3m space group are referred to as the rock salt structure, since NaCl was the first mineral proven to have this structure.

For oxide compounds, the particular crystal structure that is formed is related to the composition, the relative sizes of the atoms, and the tendency toward ionic or covalent bonding [16]. The composition of a pure crystalline material or more precisely the stoichiometry of the compound limits the possible crystal structures [13]. For example, a compound with a cation to oxygen ratio of 2:3 like Al_2O_3 cannot crystallize into the same type of structure as a compound with a cation to oxygen ratio of 1:1 like MgO [16]. The cation to oxygen ratio is constrained by the requirement that electrical neutrality be maintained. The ratio of the sizes of the cation (r_c) to the radius of the oxygen anion (r_a) also affects the types of structures that can form. As the size of the cation increases relative to oxygen, more oxygen ions can be packed around the cation center [16]. The possible coordination numbers and critical r_c/r_a ratios are given in Table 8 along with the resulting structure types [1]. Finally, the bond character also affects the crystal structure. For highly covalent crystals, the hybridization of the

Table 7 Hierarchical organization of crystal structures

Crystal system	Possible Bravais lattices	Crystal classes or point groups	Number of space groups
Cubic	P, I, F[a]	5	36
Hexagonal	P	7	27
Trigonal	P	5	25
Tetragonal	P, I	7	68
Orthorhombic	P, C, I, F	3	59
Monoclinic	P, C	3	13
Triclinic	P	2	2
7	14	32	230

[a]P primitive; C end centered; I body centered; F face centered

Table 8 Critical cation to anion radius ratios for stability various coordination environments

r_c/r_a	Coordination number	Configuration	Example
$0 \geq r_c/r_a > 0.155$	2	Linear	CO_2
$0.155 \geq r_c/r_a > 0.225$	3	Triangle	O in rutile
$0.225 \geq r_c/r_a > 0.414$	4	Tetrahedron	Wurtzite
$0.414 \geq r_c/r_a > 0.732$	6	Octahedron	Rock salt
$0.732 \geq r_c/r_a > 1.0$	8	Cubic	Fluorite
$1.0 \geq r_c/r_a$	12	Cuboctahedron	A site in Perovskite

valence shell orbitals is often the determining factor in crystal structure [16]. For example, SiC has a radius ratio of 1:6, but it crystallizes into the wurtzite structure (tetrahedral coordination) because of the strong covalent nature of the bonds [13]. A number of methods exist to predict structures including radius ratios [16], Pauling's rules [25], and Mooser-Pearson plots [13].

A majority of the important oxide ceramics fall into a few particular structure types. One omission from this review is the structure of silicates, which can be found in many ceramics [1, 26] or mineralogy [19, 20] texts. Silicate structures are composed of silicon–oxygen tetrahedral that form a variety of chain and network type structures depending on whether the tetrahedra share corners, edges, or faces. For most nonsilicate ceramics, the crystal structures are variations of either the face-centered cubic (FCC) lattice or a hexagonal close-packed (HCP) lattice with different cation and anion occupancies of the available sites [25]. Common structure names, examples of compounds with those structures, site occupancies, and coordination numbers are summarized in Tables 9 and 10 for FCC and HCP-based structures [13, 25]. The FCC-based structures are rock salt, fluorite, anti-fluorite, perovskite, and spinel. The HCP-based structures are wurtzite, rutile, and corundum.

3.3 Macroscopic Behavior

The macroscopic behavior of refractory oxides is controlled by both the bonding and crystal structure. In particular, the mechanical response and electrical behavior of materials are interpreted in terms of the symmetry of the constituent crystals using matrix or tensor algebra [27]. Other characteristics such as melting temperature and

Table 9 FCC-based crystal structures

Structure	Stoichiometry	Cation coordination	Oxygen coordination	Examples	Common characteristics
Rock salt	MO	6	6	MgO, CaO, NiO, FeO	
Fluorite	MO_2	8	4	ZrO_2, ThO_2, CeO_2	Oxygen ion conduction
Anti-fluorite	M_2O	4	8	Na_2O, Li_2O, K_2O	Fluxes, prone to hydration
Perovskite	ABO_3	$A = 12$, $B = 6$	6	$PbTiO_3$, $BaTiO_3$	High dielectric constant
Spinel	AB_2O_4	$A = 4$, $B = 6$	4	$MgAl_2O_4$, $MnFe_2O_4$	High solid solubility

Table 10 HCP-based crystal structures

Structure	Stoichiometry	Cation coordination	Oxygen coordination	Examples	Common characteristics
Wurtzite	MO	4	4	ZnO, BeO	
Rutile	MO_2	6	3	TiO_2, MnO_2	Multiple cation oxidation states
Corundum	M_2O_3	6	4	Al_2O_3, Cr_2O_3	Highly refractory

Table 11 Melting temperature [9], thermal expansion coefficient (0–1,000°C) [1], thermal conductivity (25°C) [17], elastic modulus [60], and heat capacity for some common refractory oxides

Oxide	T_m (°C)	CTE (ppm per °C)	k (W m^{-1} K^{-1})	E (GPa)	C_p (J mol^{-1} K^{-1})
Fused SiO$_2$	1,460	0.5	2	72	42.2
Quartz	1,460	10.7	13	83	56.2
TiO$_2$	1,850	7.3	8.4	290	36.9
3Al$_2$O$_3$·2SiO$_2$	1,850	5.3	6.5	220	77.1
Al$_2$O$_3$	2,020	8.8	36.2	390	78.7
MgAl$_2$O$_4$	2,135	7.6	17	239	324
ZrO$_2$	2,700	10	2.3	253	55.1
MgO	2,800	13.5	48.5	300	115.8

stability at elevated temperature are not directional and, therefore, cannot be manipulated in the same manner. However, nondirectional properties are still affected by structure in that some crystal structures are inherently more resistant to change than others. For example, structures in which some crystallographic sites are unoccupied, such as spinel, have a much higher solubility for other cations than more close-packed structures like corundum.

Phase diagrams are perhaps the most powerful tool of the materials engineer who needs to choose oxide ceramics for use at high temperature. Phase diagrams are graphical representations of the phases that are stable as a function of temperature, pressure, and composition [28, 29]. Phase diagrams can be used to determine whether a particular compound melts at a specific temperature (congruent melting), decomposes to other compounds while partially melting (incongruent melting), or reacts with another component in the system. A wide variety of phase diagrams for oxide systems are available in various compilations [9–11]. When phase diagrams are not available, behavior can be predicted with at least moderate success, using commercial programs such as FACT-SAGE [30] or using the CALPHAD methodology [31].

Considering potential applications for refractory oxides, phase diagrams also provide useful information on interactions among materials at high temperatures that might limit performance in certain gaseous atmospheres or in contact with specific liquid or solid materials. Interactions can range from the formation of low melting eutectics to reactions that form new compounds. As an example of the former, consider the effect of impurities in SiO$_2$. Pure SiO$_2$ has an equilibrium melting temperature of 1713°C [1]. All SiO$_2$, whether it is naturally occurring or prepared by other means, contains some impurities. If the presence of trace quantities of Na$_2$O are considered, a liquid phase would form at ~800°C, the SiO$_2$–Na$_2$O·2SiO$_2$ eutectic [32]. For small impurity levels, the amount of liquid increases as the amount of the second phase increases. If sufficient liquid forms to cause deformation of the component, the use temperature of silica will be reduced drastically. Eutectic liquids form when the Gibbs' energy released by mixing of the liquid components (entropic) overcomes the energy barrier (enthalpic) to melting of the unmixed solids.

Phase diagrams can also be used as an aid for material selection of oxide compounds that can be used at high temperature. Examination of diagrams (summarized in Tables 2–5) reveals that oxide compounds with melting temperatures above 1800°C are predominantly single metal oxides (e.g., Al$_2$O$_3$ or TiO$_2$) or binary oxides (e.g., $MgO \cdot Al_2O_3$ or $BaO \cdot ZrO_2$). Very few ternary oxides have high melting temperatures. The complex site occupancies and arrangements necessary to accommodate three or more cations in a single crystal structure reduce the melting temperature of ternary compounds. Upon examining ternary phase diagrams, it becomes apparent that ternary eutectic temperatures are always lower than the three binary eutectic temperatures in the corresponding binary systems. As with the binary eutectic, addition of a third component drives the eutectic temperature lower since mixing of the liquid phase components becomes more energetically favorable as the number of components increases.

It is important to distinguish between melting temperature and melting range, as the former is a fundamental property of an oxide, while the latter is a macroscopic behavior that dictates use conditions and tolerable impurity limits. Melting temperature is fairly easily understood requiring little more than observing melting of an ice cube (solid H_2O). However, only very pure substances exhibit a true melting temperature. Practical materials, except for the most pure versions (devoid of significant levels of impurity), exhibit a melting range that is defined by the macroscopic environment in which the materials exist.

In a binary combination of two oxides (e.g., alumina and silica), small additions of the second oxide result in the onset of melting at a eutectic temperature that is below the melting temperature for the pure components. For the alumina–silica system, two eutectic compositions exist depending on the overall chemistry of the mixture. For the silica-rich eutectic, all compositions between ~1 wt% and ~70 wt% alumina have an identical temperature for the onset of melting; only the amount of liquid formed will vary with composition. This temperature defines the low end of the melting range, while the temperature at which all of the material is molten (i.e., the liquidus temperature) defines the high end.

Melting range can have a profound impact on performance as liquid formation can lead to shrinkage of the refractory, reaction with the contained product, high temperature softening and flow (especially under pressure), etc. The viscosity behavior of the liquid itself is also important as highly viscous fluids behave very similarly to solids so considerably more can be present before problems occur.

4 Processing

The intrinsic properties of materials depend on bonding and crystal structure. For ceramics, the microstructure that results from the processing cycle also has a strong influence on performance. Because a majority of commercial ceramic parts are fabricated from fine powdered precursors, microstructure development during densification must be understood to control the performance of the final part. The steps in the process include powder synthesis, consolidation of powders/shaping, and densification. Powder synthesis methods range from the traditional "heat and beat" approach that uses repeated calcination and mechanical grinding steps [33] to more sophisticated reaction-based and chemical preparation methods [34]. Powder synthesis has been the subject of technical articles and reviews and will not be discussed further in this chapter. Likewise, the consolidation methods used to shape powders such as dry pressing and isostatic pressing are well documented elsewhere [35]. This section will review some key issues related to microstructure development during densification. Typical microstructures produced by solid-state sintering will be contrasted with those formed by liquid phase sintering to highlight the potential effects on performance.

4.1 Solid-State Sintering

Solid-state sintering is the preferred method used to produce fine-grained ceramics with high relative density because of process simplicity. A large variety of high purity precursor powders are commercially available with common refractory oxides such as Al_2O_3, ZrO_2, MgO, and others produced in industrially significant quantities. The process of sintering will only be briefly reviewed here since several excellent texts [36, 37] and overviews are available [38, 39]. In addition, numerous papers have been published on the sintering of specific ceramic compounds.

During solid-state sintering, porosity in powder compacts is reduced from 40 to 60 vol% in green bodies to values that can approach zero in finished parts [35]. As the porosity is removed, the volume of the part decreases while modulus and mechanical strength increase [1]. Solid-state sintering is driven by the reduction of surface free energy that occurs when high energy solid-vapor interfaces (e.g., particle surfaces) are replaced by lower energy solid-solid interfaces (e.g., grain boundaries) [37]. For academic study, the sintering process is divided into stages: (1) initial sintering, (2) intermediate sintering, and (3) final sintering [37]. These stages can be defined in terms of physical changes in the compacts such as grain size or total volume, variations in physical properties such as relative density, or differences in mechanical properties such as moduli [37]. The sintering rate, ultimate relative density, and final grain size are affected by the particular oxide chemistry that makes up the compact, its initial particle size, and the efficiency of particle packing after consolidation. In general, effective solid-state sintering is limited to powders with relatively fine (~10 μm or less) particle size.

Densification of powder compacts requires mass transport. In solid-state sintering, material is transported from the bulk or the surface of particles into pores. To overcome kinetic limitations and promote mobility of atoms, a powder compact is heated to a significant fraction of its melting temperature. Sintering temperatures for single phase oxides typically fall in the range of 0.75–0.90 of the melting temperature (T_m). For example, mullite (incongruent melting point 1890°C or 2,163 K [40]) with an initial particle size of approximately 0.2 μm can be sintered to ~98% relative density by heating to 1600°C (1,723 K or 0.80 T_m) for 2 h [41, 42]. The resulting ceramic had a final grain size of approximately 1 μm (Fig. 3) and a microstructure typical of solid state sintered, fine grained ceramics.

Sintering temperature and rate are also affected by particle size. Precursor powders with a "fine" grain size reach the same density at lower temperatures compared with "coarse" grained powders [35]. Smaller particles have a greater surface area to volume ratio and, therefore, a higher driving force for densification, which can lower the temperature required for densification [1]. In addition to precursor powder particle size, the packing of particles prior to sintering affects densification. Nonuniform particle packing can result in the formation of stable pores in fired microstructures [35]. As the pore size approaches the grain size, the driving force for pore removal approaches zero; pores that are larger than the grains are, therefore, stabilized due to a lack of driving force for removal [37]. Stable pore formation is especially problematic when

Fig. 3 Microstructure of a solid-state sintered mullite ceramic with a relative density >99%. The ceramic had equiaxed grains with a grain size of ~1 μm and no apparent glassy phase

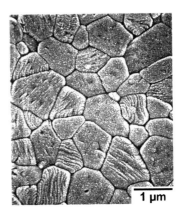

1 μm

nano-sized particles are employed because of their propensity toward formation of hard agglomerates.

The grain size of a sintered ceramic affects its performance. During solid-state sintering, grains grow during the final stage of sintering as full density is approached [43]. As with the densification process, grain growth is also driven by a reduction in surface energy; however, elimination of grain boundaries (solid-solid interfaces) in dense solids is less energetically favorable than the elimination of free surfaces (solid-vapor interfaces) in porous compacts [1]. The grain size required to achieve the performance requirements for a particular application may be smaller or larger than the grain size that would result from the optimal heat treatment. Grain growth can be altered by changing the time and temperature of the heat treatment [37]. In many cases, trace additives or dopants can be used to further modify grain growth [38]. Some dopants dissolve into the matrix altering its defect chemistry and thereby affecting material transport rates. Other dopants remain as discrete particles that affect grain growth simply by their presence in grain boundaries. The classic example of a particulate dopant that inhibits grain growth while promoting sintering is the addition of MgO to Al_2O_3. When α-Al_2O_3 is sintered at 1600°C in air, the average grain size is ~5.0 μm and a density of ~97% is achieved [44]. For the same α-Al_2O_3 doped with 250 ppm MgO sintered under the same conditions to the same density, the grain size is only ~3.5 μm [16].

4.2 Liquid Phase Sintering

Liquid phase sintering is a densification process in which a liquid phase increases the consolidation rate by facilitating particle rearrangement, enhancing transport kinetics, or both [16]. The modern practice of liquid phase sintering evolved from the vitrification of traditional ceramic ware [45, 46]. During vitrification of clay-based ceramics, heating induces the formation of a high viscosity siliceous liquid phase [1]. The liquid facilitates the dissolution of the remaining solid and the subsequent precipitation of primary mullite crystals with a needle-like morphology [47]. The fraction of liquid depends upon the particular batch composition and the firing temperature, but can be well over 50 vol% for common triaxial whitewares [45]. During vitrification, the porous powder compact undergoes physical changes (shrinkage, pore removal) and chemical reactions (conversion of meta-kaolin to mullite, glass formation). The final vitrified body is generally free of pores and contains primary mullite that is formed during vitrification, secondary mullite that is formed by precipitation from the liquid during cooling, solidified glass, plus any inert fillers such as quartz that may have been added to the batch [47]. The glassy phase serves as a bonding phase cementing the mullite crystals and fillers into a dense, strong ceramic [1]. Unlike solid-state sintering, liquid phase sintering is an effective means for densification of large grain (10 μm or greater) materials.

The modern practice of liquid phase sintering uses additives to facilitate liquid phase formation [48]. Effective liquid phase sintering minimizes liquid formation to avoid unintended deformation during densification [49]. Liquid contents as low as 3–5 vol% are possible for well-designed liquid phase sintering operations [37]. To promote densification, the liquid must form in appreciable quantities at the desired sintering temperature, it must wet the matrix, and it must be able to dissolve the matrix [49]. As with vitrification, densification during liquid phase sintering occurs by particle rearrangement and solution precipitation, which are then followed by nondensifying grain coarsening through Ostwald ripening [37]. Upon cooling, the liquid may form a glass or a crystalline phase. The solidified liquid can form a continuous film that surrounds the grains, an interpenetrating phase in the form of ligaments along grain boundaries, or an isolated phase that retreats to triple-grain junctions [1]. Liquid penetration along the grain boundaries is a function of the ratio of solid–solid interfacial energy to solid–liquid interfacial energy, which is commonly expressed as the dihedral angle [37]. To enhance performance at elevated temperatures, the amount of the residual second phase should be minimized if it is glassy upon cooling. Alternatively, some liquid phase sintering aids are designed to convert to crystalline phases that resist deformation. In either case, the resulting ceramic cannot generally operate above the temperature at which any glassy phase softens or the lower end of the melting range. In many instances, use temperatures are substantially below these limits. A representative liquid phase sintered microstructure, in this case for a mullite ceramic, has both a major phase and a solidified liquid (Fig. 4). The grain size in the liquid phase sintered ceramic is nearly an order of magnitude greater than in the solid-state sintered ceramic because of increased particle coarsening.

Liquid phase sintering processes can be designed for ceramic systems (and metallic ones for that matter) with the aid of phase diagrams [37]. The first step in designing

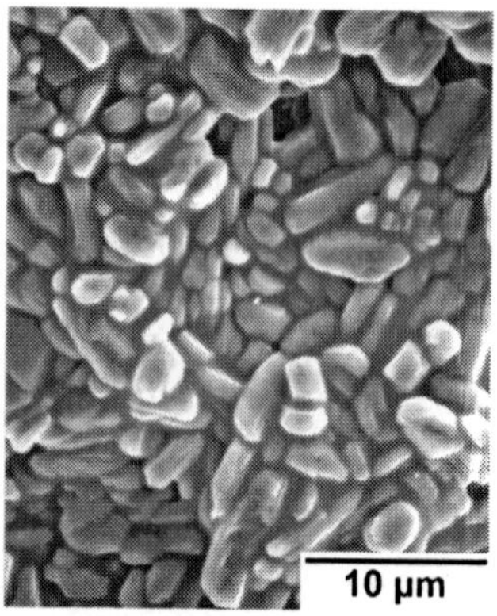

Fig. 4 Microstructure of a liquid phase sintered mullite ceramic with a relative density >99%. The ceramic had elongated grains with a grain size of >5 μm and had a residual glassy phase surrounding the mullite grains

10 μm

the liquid phase sintering process is to determine a range of compositions for the proposed additives that will promote liquid formation at the desired sintering temperature. This can be done using the appropriate binary, ternary, or higher order phase diagrams. Next, the composition of the liquid phase, after it becomes saturated with the matrix phase, can be predicted by constructing a join between the additive composition and matrix composition. Finally, the amount and composition of the phases that will be present after processing can be predicted by analyzing the cooling path for the matrix saturated liquid phase. One common example is the densification of α-Al_2O_3 with the aid of a CaO–SiO_2 glass [50]. Using the CaO–SiO_2–Al_2O_3 ternary phase diagram [51], the first choice may be to select the CaO–SiO_2 composition that results in the minimum melting temperature (64 wt% SiO_2, 36 wt% CaO, which is the binary eutectic composition that melts at 1,426°C). However, analysis of the resulting liquid phase (19 wt% CaO, 34 wt% SiO_2, 47 wt% Al_2O_3) indicates that $CaO·6Al_2O_3$, $2CaO·Al_2O_3·SiO_2$, and $CaO·Al_2O_3·2SiO_2$ will form upon final solidification by peritectic reaction at 1,380°C. For example, the Al_2O_3-saturated liquid composition lies in the $CaO·6Al_2O_3$–$2CaO·Al_2O_3·SiO_2$–$CaO·Al_2O_3·2SiO_2$ compositional triangle. The resulting ceramic would contain 91.5 wt% α-Al_2O_3 for a composition containing a 4.0 wt% sintering aid addition. To increase the α-Al_2O_3 content of the final product, the initial additive composition can be shifted to 67 wt% SiO_2 so that a liquid phase containing 19 wt% CaO, 36 wt% SiO_2, 45 wt% Al_2O_3 forms when equilibrium is reached at 1,600°C. Upon cooling, Al_2O_3, $CaO·6Al_2O_3$, and $CaO·Al_2O_3·2SiO_2$ will form by peritectic reaction at 1,495°C, increasing the resulting α-Al_2O_3 content of the final ceramic to 93.7 wt% for a composition containing a 4.0 wt% sintering aid addition. This change in composition also increases the temperature of first liquid formation by 115°C thereby allowing the ceramic to be used in higher temperature applications.

5 Properties

The critical material properties for refractory oxides are dictated by a given application. In some applications, thermal expansion and strength may be most important while in other situations melting temperature and thermal conductivity are important. In general, the most important material properties for refractory oxides include melting temperature, thermal expansion coefficient, thermal diffusivity and conductivity, elastic modulus, and heat capacity.

5.1 Melting Temperature

Melting temperature (T_m in units of °C or K) and melting range were discussed previously. The former will be the higher of the two and represents the temperature at which the phase pure oxide melts. As has been discussed, melting temperature data for selected oxides are included in Tables 2–5. A review of the literature also yields melting temperatures for many thousands of oxides that would not be classified as refractory.

During application of refractory oxides, melting range is typically more important than melting temperature. Softening point is defined as the temperature at which a material begins to deform under its own load. With phase pure oxide systems, melting temperature and softening point are equivalent; however, in most practical systems impurities are inherent. These impurities lead to low melting point eutectic formation that can lower the maximum use temperature of the oxide.

Phase equilibria diagrams yield an estimate of the softening point for a refractory oxide. Considering the binary phase diagram for the oxide and the predominant impurity, the invariant temperature (eutectic, peritectic, or monotectic) or the invariant that is closest to the refractory oxide composition indicates the lowest temperature that will result in liquid formation and, therefore, the lowest possible softening point. Although an appropriate ternary phase diagram is required, the situation is only slightly more complex when two impurities are present in significant concentrations. In that situation, initial liquid formation is defined by the invariant point for the Alkemade triangle between the refractory oxide and the two impurities. For example, pure SiO_2 has an equilibrium melting temperature of 1713°C. The addition of Na_2O reduces the eutectic temperature to ~800°C at the SiO_2-rich end of the diagram. Adding a third oxide, K_2O, reduces the eutectic temperature to ~540°C at the SiO_2-rich end of the diagram.

5.2 Thermal Expansion

Thermal expansion is the change in specific volume of a material as it is heated. The linear coefficient of thermal expansion (α with units of inverse temperature) can be expressed as the change in length of an object, normalized by its original length, for a given temperature change (3):

$$\alpha = \left(\frac{\Delta L}{L} \right)\left(\frac{1}{\Delta T} \right)$$

(3)

where ΔL is the change in length for a given temperature change (m), L is the original length (m), and ΔT is the change in temperature (K).

In general, all materials have a positive thermal expansion coefficient; that is they increase in volume when heated. Thermal expansion results from thermal excitation of the atoms that compose the material [16]. At absolute zero, atoms are at rest at their equilibrium positions (i.e., at r_0 in Fig. 1). As they are heated, thermal energy causes the atoms to vibrate around their equilibrium positions. The amplitude of vibration increases as heating is continued. Asymmetry in the shape of the potential well causes the average interatomic distance to increase as temperature increases, leading to an overall increase in volume [15].

The importance of considering thermal expansion cannot be underestimated. Ignoring thermal expansion or incorrectly accounting for thermal dilations can have serious consequences. Consider a vessel that is ~3 m (~10 ft) in diameter insulated with a zirconia refractory. Assuming a linear coefficient of thermal expansion of 13 ppm K^{-1}, heating the lining from room temperature to 1600°C the inside surface of the lining would grow by about 5 cm (~2 in.). To compensate for expansion in large systems, it is common to leave expansion joints spaced at regular intervals. When temperature is increased, the refractory material will shift into the open space preventing potential problems.

5.3 Thermal Diffusivity/Conductivity

Thermal conductivity (k with units W m^{-1} K^{-1}) describes the ability of a material to transport thermal energy because of temperature gradient. Steady-state thermal conductivity is a constant of proportionality between the heat flux (time rate of heat flow per unit area) through a solid and the imposed temperature gradient as described by (4) [52]:

$$\frac{Q}{A} = k \frac{\Delta T}{x}, \qquad (4)$$

where Q is the heat flow (J s^{-1} or W), A is cross sectional area (m^2), k is thermal conductivity (W m^{-1} K^{-1}), ΔT is temperature gradient (K), and x is distance (m).

In electrically insulating solids, heat is transferred in the form of elastic waves or phonons [1]. Anything that affects the propagation of the phonons through the solid affects the thermal conductivity of the solid. In a pure crystalline ceramic, the intrinsic thermal conductivity is limited by the energy dissipated during phonon–phonon collisions or so-called Umklapp processes [15]. Commonly, the intrinsic thermal conductivity of solids is described by (5).

$$k = \frac{1}{3} Cvl, \qquad (5)$$

where C isthe heat capacity per unit volume (J m^{-3} K^{-1}), v is the phonon velocity (m s^{-1}), and l is phonon mean free path (m).

Phonon velocity and mean free path are difficult to measure accurately in polycrystalline materials, so (5) is normally restricted to theoretical predictions. The values of thermal conductivity observed in polycrystalline ceramics are often significantly less than the intrinsic values predicted or those measured for single crystals. Specimen characteristics such as temperature, impurities, grain size, porosity, and preferred orientation affect the phonon mean free path thereby changing thermal conductivity [1]. Though not an oxide, this effect is pronounced in aluminum nitride. The intrinsic thermal conductivity of AlN is 280 W m^{-1} K^{-1} [16], but thermal conductivities in the range of 50–150 W m^{-1} K^{-1} are often observed in sintered materials because of the presence of grain boundaries and second phases [53].

The thermal conductivity of large grained (100 μm or more) ceramics can be determined by direct measurement techniques described in ASTM standards C 201-93 (Standard Test Method for Thermal Conductivity of Refractories), C 1113-99 (Standard Test Method for Thermal Conductivity of Refractories by Hot Wire), and E 1225-99 (Standard Test Method for Thermal Conductivity of Solids by Means of the Guarded-Comparative-Longitudinal Heat Flow Technique). These methods lend themselves to quality control-type assessment of the thermal conductivity of macroscopic parts in standard shapes (e.g., 9 in. straight brick or monolithic materials cast to specific dimensions). The sizes prescribed by these standards insure that the specimen thickness is sufficient to reflect the effects of grain boundaries, pores, and other specimen characteristics. The relative error of the techniques ranges from ~10 to ~30% depending on the material and technique. This degree of precision is normally sufficient for material selection and design calculations. Some common design considerations influenced by thermal conductivity include cold-face temperature, interface temperatures between working lining and insulating lining materials, heat loss, and estimating the required thickness for each component in a system. For example, the heat flux through a refractory material can be calculated using (4). Assuming a characteristic thermal conductivity for an insulating firebrick of $0.25\,W\,m^{-1}\,K^{-1}$ at the mean temperature of the wall, a heat flow of $5000\,J\,s^{-1}$ would be predicted per square meter of area for a hot face temperature of $1200°C$ ($1473\,K$), a cold face temperature of $200°C$ ($473\,K$), and a wall thickness of $5\,cm$ ($0.05\,m$).

For dense specimens of fine-grained (less than $100\,\mu m$) technical ceramics, the thermal conductivity can be determined with greater precision using an indirect method by which thermal diffusivity (α with units of $m^2\,s^{-1}$) is measured and then converted to thermal conductivity. For small specimens, precise control of heat flow and accurate determination of small temperature gradients can be difficult, leading to unacceptably large error in the direct measurement of thermal conductivity of small specimens. As a consequence, determination of thermal diffusivity by impulse heating of thin specimens followed by conversion to thermal conductivity has evolved as the preferred measurement technique [54, 55]. The technique is described in ASTM standard E1461–01 (Standard Test Method of Thermal Diffusivity by the Flash Method). Measured thermal diffusivity is used to calculate thermal conductivity using (6):

$$\alpha = \frac{k}{C_p \rho} \tag{6}$$

where α is thermal diffusivity ($m^2\,s^{-1}$), k is thermal conductivity ($W\,(m^{-1}\,K)^{-1}$), C_p is heat capacity ($J\,kg^{-1}\,K^{-1}$), ρ is density ($kg\,m^{-3}$).

5.4 Elastic Modulus/Strength

Elastic modulus or Young's modulus (E with units of GPa) describes the response of a linear elastic material to an applied mechanical load [16]. Elastic modulus relates the applied load to the resulting strain as expressed by Hooke's law (7).

$$\sigma = E\varepsilon \tag{7}$$

where σ is the applied stress (GPa) and ε is the strain (no units).

Under an applied load, deformation of the solid requires that the atoms be moved closer together (compressive load) or farther apart (tensile load). As such, dimensional changes are related to the strength of the bonds among the atoms [8]. When the component ions of a material have high bond strengths, the material typically displays high elastic modulus and low coefficient of thermal expansion. For example, SiC has a high bond strength giving sintered α-SiC a coefficient of thermal expansion of $4.02\,ppm\,K^{-1}$ and a modulus of $410\,GPa$ [56]. Conversely, NaCl has a low bond strength resulting in a coefficient of thermal expansion of $11.0\,ppm\,K^{-1}$ and a modulus of $44\,GPa$ [16, 57]. Modulus can be measured using either acoustic methods (ASTM E 1876 Standard Test Method of Dynamics Young's Modulus, Shear Modulus, and Poisson's Ratio by Impulse Excitation of Vibration) or by directly measuring displacement as a function of an applied load using a deflectometer.

Although modulus is a fundamental material property related to the strength of the chemical bonds among atoms, the measured strength (σ with units of MPa) is affected by specimen characteristics, the testing environment, the type of test performed, and other factors. The theoretical tensile strength can be estimated as the stress required to break the chemical bonds among the atoms of a solid [57]. However, brittle materials fail at applied stresses two or more orders of magnitude below the theoretical strength values due to stress concentration around physical features of the solids such as pores, defects, grain boundaries, and edges. The Griffith criterion is often used to relate the fundamental material properties such as modulus to observed strength using specimen characteristics such as flaw size [1], although the predictions are qualitative at best. The Griffith criterion can be used to understand the statistical nature of fracture of brittle materials if the distribution of flaw sizes within a given specimen is considered [16]. Strength can be measured in many different manners ranging from compression (ASTM C1424 Standard Test Method for Monotonic Compressive Strength of Advanced Ceramics at Ambient Temperatures) to tension (ASTM C1273 Standard Test Method for Tensile Strength of Monolithic Advanced Ceramics at Ambient Temperatures). The strength of advanced ceramics is most often measured using relatively small specimens in three- or four-point bending, which determines the so-called flexural strength (ASTM C1161 Standard Test Method for Flexural Strength of Advanced Ceramics at Ambient Temperature). Because traditional refractory materials have grain sizes that approach or exceed the size of the specimens used for testing of advanced ceramics, strengths must be determined using alternate methods (ASTM C133 Standard Test Methods for Cold Crushing Strength and Modulus of Rupture of Refractories) that accommodate course grain specimens.

5.5 Heat Capacity

Heat capacity (C_p with units such as J mol^{-1} K^{-1} or J kg^{-1} K^{-1}) is defined as the quantity of thermal energy required to raise the temperature of a substance one degree [58]. In practice, heat capacity and the term specific heat are used almost interchangeably. For ionic solids, atoms can be modeled as centers of mass that can vibrate independently in three dimensions [59]. The vibrational energy of the atoms increases as thermal energy is added to the system. The heat capacity of all solids approaches $3N_A k$ with temperature (where N_A is Avagadro's number and k is the Boltzmann constant) or $3R$ (where R is the ideal gas constant) per mole of atoms, the familiar Dulong-Petit law. At low temperature, the models of Einstein and Debye can be used to estimate heat capacity [15]. In practice, heat capacity in terms of energy and mass is more useful and is compiled as a function of temperature in any number of reference books [2, 3, 60]. Experimentally, heat capacity can be determined using such heat flow techniques as differential scanning calorimetry.

Heat capacity is important because it regulates the amount of energy required to raise the temperature of a thermal load (e.g., ware to be fired plus kiln furniture). Such data can be used to compute furnace efficiency, which is the ratio of fuel usage to the thermal energy requirement of a process. In addition, knowledge of heat capacities of products, kiln furniture, and refractories is essential for good furnace design.

6 Summary

Refractory oxides are an important class of materials that enable processes to exploit extreme environments. A wide variety of unary, binary, and ternary oxides can be considered refractory, based on their melting temperatures. Refractory oxides are generally prepared from powdered precursors using standard ceramic forming techniques such as casting, pressing, or extrusion, and subsequently sintered to achieve final density. In addition to chemical compatibility, the physical properties of refractory oxides such as thermal expansion coefficient, thermal conductivity, modulus of elasticity, and heat capacity must be considered when selecting an oxide for a specific application.

References

1. W.D. Kingery, H.K. Bowen, and D.R. Uhlmann, *Introduction to Ceramics*, 2nd edn., Wiley, New York, 1976.
2. M.W. Chase, Jr., *NIST-JANAF Thermochemical Tables*, 4th edn., American Institute of Physics, Woodbury, NY, 1998.
3. R.C. Weast and M.J. Astle (eds.), *Handbook of Chemistry and Physics*, 62nd edn., CRC Press, Boca Raton, 1981.
4. H. Ries, *Clays, Their Occurrence, Properties, and Uses*, 3rd edn., Wiley, New York, 1927.
5. F.H. Norton, *Refractories*, 2nd edn., McGraw-Hill, New York, 1942.
6. Harbison-Walker Refractories, *HW Handbook of Refractory Practice*, 2nd edn., Harbison-Walker Refractories, Pittsburgh, 1980.
7. TARJ, *Refractories Handbook*, The Technical Association of Refractories, Japan, Tokyo, 1998.
8. W.D. Kingery, *Introduction to Ceramics*, Wiley, New York, 1960.
9. The American Ceramic Society, *Phase Diagrams for Ceramists*, Vols. 1–13, The American Ceramic Society, Columbus/Westerville, OH, 1964–2001.
10. R.S. Roth, The American Ceramic Society (eds.), *Phase Diagrams for Electronic Ceramics I: Dielectric Ti, Nb, and Ta Oxide Systems*, The American Ceramic Society, OH, 2003.
11. H.M. Ondik and H.F. McMurdie (eds.), *Phase Diagrams for Zirconium and Zirconia Systems*, The American Ceramic Society, OH, 1998.
12. L. Pauling, *The Nature of the Chemical Bond*, 3rd edn., Cornell University Press, Ithaca, NY, 1960.
13. A.R. West, *Basic Solid State Chemistry*, 2nd edn., Wiley, Chichester, 1999.
14. H.L. Brown and H.E. LeMay, Jr., *Chemistry: The Central Science*, 2nd edn., Prentice-Hall, NJ, 1981.
15. C. Kittel, *Introduction to Solid State Physics*, 6th edn., Wiley, New York, 1986.
16. M. Barsoum, *Fundamentals of Ceramics*, McGraw Hill, New York, 1997.
17. Y.S. Touloukian, R.K. Kirby, R.E. Taylor, and T.Y.R. Lee, *Thermophysical Properties of Matter Volume 13: Thermal Expansion, Non Metallic Solids*, IFI Plenum, New York, 1977.
18. R.C. Evans, *An Introduction to Crystal Chemistry*, 2nd edn., University Press, Cambridge, 1964.
19. C. Klein and C.S. Hurlbut, Jr., *Manual of Mineralogy*, 21st edn., Wiley, New York, 1993.
20. B.D. Cullity and S.R. Stock, *Elements of X-Ray Diffraction*, 3rd edn., Prentice-Hall, NJ, 2001, pp. 31–88.
21. R. Jenkins and R.L. Snyder, *Introduction to X-Ray Powder Diffractometry*, Wiley, New York, 1996, pp. 23–46.
22. F.D. Bloss, *Crystallography and Crystal Chemistry; An Introduction*, Holt, Rinehart, and Winston, New York, 1971.
23. W.D. Nesse, *Introduction to Optical Mineralogy*, Oxford University Press, New York, 1986.

24. Y.-M. Chiang, D. Birnie, III, and W.D. Kingery, *Physical Ceramics: Principles for Ceramic Science and Engineering*, Wiley, New York, 1997, pp. 13–19.
25. F.H. Norton, *Fine Ceramics*, McGraw-Hill, New York, 1970, p. 40.
26. J.F. Nye, *Physical Properties of Crystals: Their Representation by Tensors and Matrices*, Clarendon Press, Oxford, 1985.
27. C.G. Bergeron and S.H. Risbud, *Introduction to Phase Equilibria in Ceramics*, The American Ceramic Society, Columbus, OH, 1984.
28. F.A. Hummel, *Introduction to Phase Equilibria in Ceramic Systems*, Marcel Dekker, New York, 1984.
29. Available from the Centre for Research in Computational Thermochemistry at Ecolé Polytechnique, Box 6079, Station Downtown, Montreal, Quebec, Canada.
30. CALPHAD: Computer Coupling of Phase Diagrams and Thermochemistry, The International Research Journal for Calculation of Phase Diagrams, published quarterly by Elsevier.
31. E.M. Levin, C.R. Robbins, and H.F. McMurdie, *Phase Diagrams for Ceramists*, Vol. 1, The American Ceramic Society, Columbus, OH, 1964, Figure 192.
32. G.Y. Onoda and L.L. Hench (eds.), *Ceramic Processing Before Firing*, Wiley, New York, 1978.
33. C.J. Brinker, D.E. Clark, and D.R. Ulrich (eds.), *Better Ceramics Through Chemistry*, Materials Research Society Proceedings Volume 32, North Holland, New York, 1984.
34. J.S. Reed, *Principles of Ceramics Processing*, Wiley, New York, 1995.
35. R.M. German, *Sintering Theory and Practice*, Wiley, New York, 1996.
36. M.N. Rahaman, *Ceramic Processing and Sintering*, Marcel Dekker, New York, 1995.
37. M.F. Yan, Solid state sintering, in *Ceramics and Glasses: Engineered Materials Handbook*, Vol. 4, ASM International, Materials Park, OH, 1991, pp. 270–284.
38. M.P. Harmer, Science of sintering as related to ceramic powder processing, in *Ceramic Powder Science II, Ceramic Transactions*, Vol. 1, G.L. Messing, E.R. Fuller, Jr., and H. Hausner (eds.), The American Ceramic Society, Westerville, OH, 1988, pp. 824–839.
39. F.J. Klug, S. Prochazka, and R.H. Doremus, Alumina-silica phase diagram in the mullite region, *J. Ame. Ceram. Soc.* **70**(10) 750–759 (1987).
40. W.G. Fahrenholtz, *Particle size and mixing effects on the crystallization and densification of sol-gel mullite*, Ph.D. Thesis, University of New Mexico, 1992.
41. W.G. Fahrenholtz, D.M. Smith, and J. Cesarano III, Effect of precursor particle size on the densification and crystallization behavior of mullite, *J. Am. Ceram. Soc.* **76**(2), 433–437 (1993).
42. R.J. Brook, Controlled grain growth, in *Ceramic Fabrication Processes: Treatise on Materials Science and Technology*, Vol. 9, F.F.Y. Wang (ed.), Academic Press, New York, 1976, pp. 331–364.
43. K.A. Berry and M.P. Harmer, Effect of MgO solute on microstructure development in Al_2O_3, *J. Am. Ceram Soc.* **69**(2), 143–149 (1986).
44. F.H. Norton, *Fine Ceramics*, McGraw-Hill, New York, 1970, pp. 258–280.
45. G.W. Brindley and M. Nakahira, The kaolinite-mullite reaction series: I, a survey of outstanding problems; II, metakaolin; and III, the high temperature phases, *J. Am. Ceram. Soc.* **42**(7), 311–324 (1959).
46. Y. Iqbal and W.E. Lee, Microstructural evolution in triaxial porcelain, *J. Am. Ceram. Soc.* **83**(12), 3121–3127 (2000).
47. R.M. German, *Liquid Phase Sintering*, Plenum, New York, 1985.
48. O.-H. Kwon, Liquid phase sintering, in *Ceramics and Glasses: Engineered Materials Handbook* Vol. 4, ASM International, Materials Park, OH, 1991, pp. 285–290.
49. O.-H. Kwon and G.L. Messing, Kinetic analysis of solution-precipitation during liquid-phase sintering of alumina, *J. Am. Ceram. Soc.* **73**(2), 275–281 (1990).
50. E.M. Levin, C.R. Robbins, and H.F. McMurdie (eds.), *Phase Diagrams for Ceramists*, Vol. 1, The American Ceramic Society, Columbus, OH, 1964, Figure 630.
51. R.B. Bird, W.E. Stewart, and E.N. Lightfoot, *Transport Phenomena*, Chap. 8, Wiley, New York, 1960.
52. S. Hampshire, Engineering properties of nitrides, in *Ceramics and Glass: Engineered Materials Handbook*, Vol. 4, ASM International, Materials Park, OH, 1991, pp. 812–820.
53. W.J. Parker, R.J. Jenkins, C.P. Butler, and G.L. Abbot, Flash method of determining thermal diffusivity, heat capacity, and thermal conductivity, *J. Appl. Phy.* **32**(9), 1979–1984 (1961).
54. T. Log and T.B. Jackson, Simple and inexpensive flash technique for determining thermal diffusivity of ceramics, *J. Am. Ceram. Soc.* **74**(5), 941–944 (1991).

55. Peter T.B. Shaffer, Engineering Properties of Carbides, in *Ceramics and Glass: Engineered Materials Handbook*, Vol. 4, ASM International, Materials Park, OH, 1991, pp. 804–811.
56. D.W. Richerson, *Modern Ceramic Engineering*, Marcel Dekker, New York, 1992.
57. D.V Ragone, *Thermodynamics of Materials*, Vol. 1, Wiley, New York, 1995, p. 12.
58. H.M. Rosenberg, *The Solid State*, 2nd edn., Chap. 5, Oxford University Press, New York, 1978.
59. Thermochemical and Physical Property database, Version 2.2, ESM software.

Chapter 7
Clays

William G. Fahrenholtz

Abstract Clays are ubiquitous constituents of the Earth's crust that serve as raw materials for traditional ceramics. Mineralogically, clays are phyllosilicates or layered aluminosilicates. Bonding is strong within layers, but weak between layers, allowing clays to break into micrometer-sized particles. When mixed with water, clays develop plasticity and can be shaped easily and reproducibly. When heated, clays undergo a series of reactions that ultimately produce crystalline mullite and a silica-rich amorphous phase. Beyond the structure and properties of clays, the science that developed to understand traditional ceramics continues to serve as the framework for the study of advanced ceramics.

1 Introduction and Historic Overview

Products such as bricks, whitewares, cements, glasses, and alumina are considered traditional ceramics because they are derived from either (1) crude minerals taken directly from deposits or (2) refined minerals that have undergone beneficiation to remove mineral impurities and control physical characteristics [1]. Most traditional ceramics are fabricated using substantial amounts of clay. Clays are distinguished from other naturally occurring raw materials by their development of plasticity when mixed with water [2]. As a common mineral constituent of the Earth's crust, clays have been used to fabricate useful objects for countless generations, with earthenware ceramics dating back to at least 5000 B.C. [3]. Clay-based ceramic objects were used by virtually all pre-historic cultures for practical, decorative, and ceremonial purposes. Analysis of shards from these objects is our primary means of gathering information on these civilizations. The hard porcelains produced by the ancient Chinese (~575 A.D. more than 100 years before their European counterparts) stand as the ultimate achievement in the field of ceramics prior to the industrial revolution [4,5]. Clay minerals continue to be widely utilized in the production of traditional ceramics and other products due to their ubiquity and low cost combined with properties that include plasticity during forming, rigidity after drying, and durability after firing [6].

For much of the twentieth century, the ceramics industry centered on the utilization of clays and other silicate minerals. Ceramic engineering educational programs and organizations such as the American Ceramic Society were founded to serve industries

J.F. Shackelford and R.H. Doremus (eds.), *Ceramic and Glass Materials:*
Structure, Properties and Processing.
© Springer 2008

based on the utilization of silicate or aluminosilicate minerals [7]. As clay-based traditional ceramics became commodity items in the middle and latter portions of the twentieth century, the focus of educational programs and industrial development shifted away from mineral utilization and toward advanced ceramics, which include phase-pure oxides, electronic materials, and non-oxide ceramics. The raw materials for these products are classified as industrial inorganic chemicals because they have been chemically processed to improve purity compared with the crude or refined minerals used to produce traditional ceramics [1]. Despite the shift in focus away from traditional ceramics, the production of clays has not fallen significantly over the past 30 years (Fig. 1). At a current average cost of more than $30 per ton (Fig. 2), clay production was a $1.3 billion

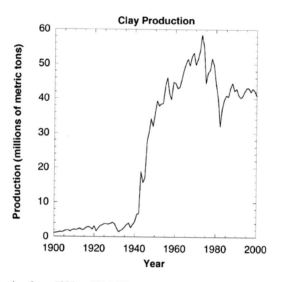

Fig. 1   Clay production from 1900 to 2002 [8]

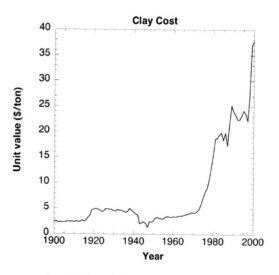

Fig. 2   Cost of clay per ton from 1900 to 2002 [8]

industry in 2002, based on 40.7 million metric tons [8]. Traditional ceramics still account for a significant fraction of the total industry with production in the nonmetallic minerals sector that produced approximately $95 billion in goods during 2001 [9].

2 Structure, Formation Mechanisms, Types of Deposits, and Use of Clays

This section reviews several of the methods that are used to categorize clays. First, the structure of clay minerals will be discussed. Next, the mechanism of formation for kaolinite will be reviewed followed by a description of the types of deposits in which clays are found. The section will end with a description of the types of clays used in the ceramics industry.

2.1 Structure of Clay Minerals

The outermost layer of our planet, the crust, contains the accessible mineral wealth of the planet. The eight most abundant elements in the crust (Table 1) make up 98.5% of the mass of the crust [10]. The most common metal, silicon, is never found in its elemental form in nature. Instead, silicon is combined in silicate minerals, which make up more than 90% of the mass of the Earth's crust [11]. Depending on the composition and formation conditions, silicate minerals have structures that range from individual clusters (orthosilicates) to three-dimensional networks (tecto-silicates) [11]. These minerals can be contained in relatively pure single mineral deposits or, more commonly, in rocks such as granite that are made up of one or more mineral species.

The term clay refers to fine-grained aluminosilicates that have a platy habit and become plastic when mixed with water [11]. Dozens of minerals fall under the classification of clays and a single clay deposit can contain a variety of individual clay minerals along with impurities. Clay minerals are classified as phyllosilicates because of their layered structure [12]. The most common clay mineral is kaolinite, although others such as talc, montmorillonite, and vermiculite are also abundant. Each of the

Table 1 Chemical composition of the Earth's crust

Element	Percent by Weight
O	50
Si	26
Al	7.5
Fe	4.7
Ca	3.4
Na	2.6
K	2.4
Mg	1.9
All others	1.5

clay minerals is composed of a unique combination of layers that are made up of either tetrahedral or octahedral structural units that form sheets [13]. Tetrahedral sheets are made up of oriented corner-shared Si–O tetrahedra (Fig. 3) [14]. Each tetrahedron shares three of its corners with three adjacent tetrahedra, resulting in a structural formula of $(Si_2O_5)_n$ for the sheet [15]. Likewise, octahedral sheets are composed of Al bonded to O or OH anions, resulting in an effective chemical formula of $AlO(OH)_2$ [15,16]. The structure of this sheet is shown in Fig. 4 [14]. The simplest clay mineral, kaolinite, is produced when each of the Si–O tetrahedra in the tetrahedral sheet shares an oxygen with an Al–O/OH octahedron from the octahedral sheet, shown as a perspective drawing in Fig. 5. The repeat unit or layer in the resulting structure is

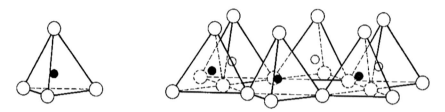

Fig. 3 A single Si–O tetrahedron and the structure of the tetrahedral sheet (Reproduced by permission of the McGraw-Hill companies from R.E. Grim, *Applied Clay Mineralogy*, McGraw-Hill, New York, 1962) [14]

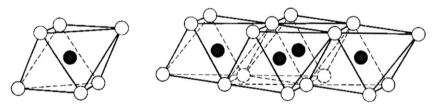

Fig. 4 A single Al–O octahedron and the structure of the octahedral sheet (Reproduced by permission of the McGraw-Hill companies from R.E. Grim, *Applied Clay Mineralogy*, McGraw-Hill, New York, 1962) [14]

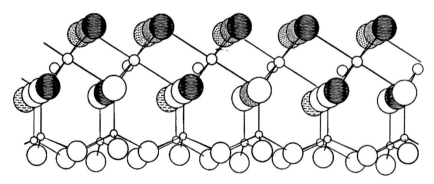

Fig. 5 Perspective drawing of the kaolinite structure taken from Brindley (Reproduced by permission of MIT Press from G.W. Brindley, "Ion Exchange in Clay Minerals," in *Ceramic Fabrication Processes*, Ed. by W.D. Kingery, John Wiley, New York, 1958, pp. 7–23) [13]

composed of alternating octahedral and tetrahedral sheets. Bonding within each repeat unit is covalent, making the layers strong. In contrast, the bonding between repeat units is relatively weak, allowing the layers to separate when placed in an excess of water or under a mechanical load. The chemical formula for kaolinite, as determined by site occupancy and charge neutrality requirements, is $Al_2Si_2O_5(OH)_4$, which is commonly expressed as the mineral formula $Al_2O_3 \cdot 2SiO_2 \cdot 2H_2O$. The structure and properties of kaolinite are summarized in Table 2. The repeat units for clay minerals other than kaolinite are produced by altering the stacking order of the octahedral and tetrahedral sheets or by isomorphous substitution of cations such as Mg^{2+} and Fe^{3+} into the octahedral sheets [17].

Conceptually, the next simplest clay mineral is pyrophyllite, which is produced by attaching tetrahedral sheets above and below an octahedral layer (Fig. 6), compared with just one octahedral sheet for kaolin [15]. The resulting chemical composition of pyrophyllite is $Al_2Si_4O_{10}(OH)_2$, which is equivalent to the mineral formula $Al_2O_3 \cdot 4SiO_2 \cdot H_2O$. The structure and properties of pyrophyllite are summarized in Table 2.

Table 2 Composition and crystallography of common clay minerals

	Kaolinite	Pyrophyllite	Mica (Muscovite)
Chemical formula	$Al_2Si_2O_5(OH)_4$	$Al_2Si_4O_{10}(OH)_2$	$KAl_3Si_3O_{10}(OH)_2$
Mineral formula	$Al_2O_3 \cdot 2SiO_2 \cdot 2H_2O$	$Al_2O_3 \cdot 4SiO_2 \cdot H_2O$	$K_2O \cdot 3Al_2O_3 \cdot 6SiO_2 \cdot 2H_2O$
Crystal class	Triclinic	Monoclinic	Monoclinic
Space group	$P\bar{1}$	$C2/c$	$C2/c$
Density	2.6 g cm^{-3}	2.8 g cm^{-3}	2.8 g cm^{-3}
c-Lattice parameter	7.2 Å	18.6 Å	20.1 Å

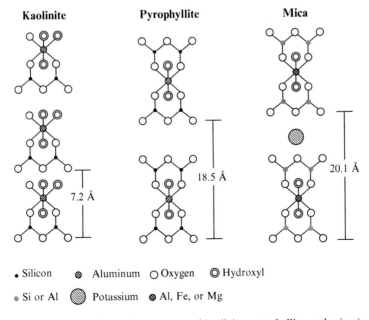

Fig. 6 Schematic representation of the structure of koalinite, pyrophyllite, and mica (muscovite) after Brindley [13]

More complex clay minerals are produced when Mg^{2+} or Fe^{3+} substitute onto the octahedral Al^{3+} sites in either the kaolinite or the pyrophyllite structures [17]. Along with the substitution onto the octahedral sites, Al^{3+} can substitute onto the tetrahedral sites. These substitutions produce a net negative charge on the structural units, which, in turn, can be compensated by alkali (Na^+, K^+) or alkaline earth (Ca^{2+}, Mg^{2+}) cations that attach to the structure either between the layers of the structural units or within the relatively large open space inside the Si–O tetrahedra [13]. Families of clay minerals that contain isomorphous substitutions on Al^{3+} and/or Si^{4+} sites are micas and chlorites. The structure of a potassium compensated mica-type mineral is shown in Fig. 6. The charge-compensating cations in these clays are relatively mobile, giving some clays significant cation exchange capacity [15]. In addition to the distinctly different minerals produced by altering the arrangement of the structural units or by substituting cations into the structure, some clays are susceptible to hydration of the interlayer cations, which can cause swelling in the c-direction. An almost infinite number of clay minerals can be conceived by varying site occupancy and layer orders. These structures can be complex and difficult to determine by experimental methods such as X-ray diffraction. Further complication arises due to the fact that some clays are made up of layers with different structural units (e.g., a random sequence of pure or partially substituted pyrophyllite- and kaolinite-type layers).

An additional structural variant for clay minerals is the chlorite-type structure. Chlorites are similar to the pyrophyllite-type structures with two tetrahedral sheets and an octahedral sheet making up each layer. Instead of alkali or alkaline earth interlayer cations, chlorites contain a brucite (Al–Mg hydroxide) layer between successive pyrophyllite-type layers [18].

The major mineralogical classifications associated with clays are summarized in Fig. 7 [18]. Fortunately as ceramists, we are more concerned with the properties of clays than their mineralogy and most often we classify them by use.

2.2   Formation Mechanism for Kaolinite

Geologically, clay minerals can be classified based on the conditions under which they form. Clay minerals can form at or near the surface of the Earth by the action of liquid water that originates either on the surface or ground water that is percolating toward the surface [6]. Clay minerals can also form under pressure at greater depths due to the action of heated (~100–450°C) liquid-water or liquid-vapor mixtures [19]. For both formation condition, three different mechanisms have been proposed for the conversion of aluminosilicate minerals to clays: (1) the direct reaction with water, (2) dissolution and removal of carbonate minerals, leaving insoluble clay impurities behind, or (3) the action of water on compacted shale sediments [6]. Only the first of these mechanisms will be discussed as it pertains to formation of clays at or near the surface of the Earth, since this combination has produced the largest volumes of industrially relevant clays. In addition, only the reaction of the most common group of minerals, the feldspars, will be considered, but it is recognized that many other minerals convert to clays. To understand the source of impurities in clays, which will be discussed in the next section, the mineralogy of the rocks that serve as the aluminosilicate source are discussed in this section.

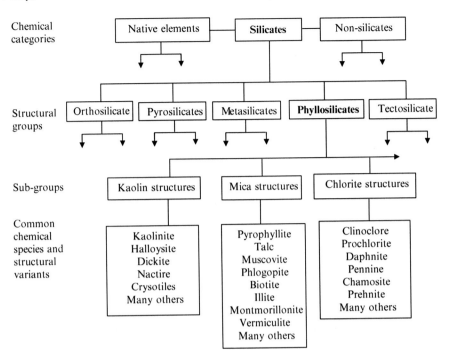

| Chemical categories | Native elements | — | **Silicates** | — | Non-silicates | | |

Structural groups: Orthosilicate | Pyrosilicates | Metasilicates | **Phyllosilicates** | Tectosilicate

Sub-groups: Kaolin structures | Mica structures | Chlorite structures

Common chemical species and structural variants:

Kaolin structures	Mica structures	Chlorite structures
Kaolinite	Pyrophyllite	Clinoclore
Halloysite	Talc	Prochlorite
Dickite	Muscovite	Daphnite
Nactire	Phlogopite	Pennine
Crysotiles	Biotite	Chamosite
Many others	Illite	Prehnite
	Montmorillonite	Many others
	Vermiculite	
	Many others	

Fig. 7 Mineralogical classifications associated with clay minerals [12,18,22]

Feldspars are common aluminosilicate minerals that are present in many different igneous rocks including granites and rhyolites [11]. When exposed, these rocks are susceptible to physical and chemical attack. Water, along with the sun, plant roots, and other forces physically attack rock formations causing crevice formation and fracture [3]. Water also attacks rocks chemically. Over time, anhydrous aluminosilicate compounds such as those present in igneous minerals react with water to form hydrated species [20]. The classic chemical reaction for clay formation involves the decomposition of potash feldspar due to the action of water-containing dissolved CO_2 to form kaolinite (insoluble) and soluble ionic species (Reaction 1) [14].

$$K_2O \cdot Al_2O_3 \cdot 6SiO_{2(s)} + 2H_2O + CO_{2(aq)} \rightarrow Al_2O_3 \cdot 2SiO_2 \cdot 2H_2O_{(s)}$$
$$+ 4SiO_{2(s)} + K_2CO_{3(aq)} \tag{1}$$

In nature, the formation of clays is more complex. One complexity is due to the variable composition of feldspar and the other is due to minerals that can react to form clays [11]. Even when only feldspars are considered, the composition can vary significantly among the end-members of the system, which are orthoclase ($K_2O \cdot Al_2O_3 \cdot 6SiO_2$), albite ($Na_2O \cdot Al_2O_3 \cdot 6SiO_2$), and anorthite ($CaO \cdot Al_2O_3 \cdot 2SiO_2$) [11]. The different feldspars along with many other aluminosilicate minerals can undergo conversion to kaolinite. Another complexity is due to the fact that feldspars and other aluminosilicates are present in nearly all igneous rocks [12]. Most often, the formation of clay is considered in the context of the decomposition of granite, a rock that contains feldspar, quartz, and mica [20]. Quartz and mica, which form due to incomplete decomposition of feldspar, are much more resistant to hydration than feldspar and are often left unaltered by the formation of clays from granite. As a result, quartz and mica are common impurities in primary clays.

2.3   Types of Clay Deposits

In nature, clays can be found either in the same location where they were formed or they can be found in a location where they were transported after formation. Clay deposits that are found where they were formed are referred to as primary or residual deposits. Clays that have been transported after formation are said to be in secondary or sedimentary deposits. The discussion in this section will be limited to kaolinite, but will be expanded to other types of clays of significance to the ceramics industry in the following section.

2.3.1   Primary Clays

Primary clay deposits are formed when a rock formation is chemically attacked by water. The size and shape of the deposit depends on the size and shape of the parent rock [6]. The mineral constituents and impurities of a primary clay deposit are also determined by the composition of the parent rock, the degree of completion of the reaction, the impurities that are removed by solution during or after reaction, and the impurities brought in during or after formation [3]. The residual clay deposits formed by conversion of feldspar almost always contains silica (quartz) and mica as major mineral impurities. The soluble cations such as potassium, sodium, and calcium are dissolved and removed during or after conversion [2]. Most primary deposits contain a high proportion of impurity phases, with typical clay contents ranging from 10 to 40% by volume [21]. However, primary deposits tend to be low in iron-bearing impurities (reported subsequently as Fe_2O_3), TiO_2, and organics. The major mineral impurities can be removed by beneficiation techniques such as air or water flotation to yield usable clay, while removal of other impurities may require more involved treatment processes [1]. Though not mineralogically correct, clays that are white in color and have minimal iron-based impurities are often referred to as "kaolin," regardless of the crystalline phases present. To avoid confusion, the term "china clay" will be used for iron-free, white burning clays in this article. Most of the commercially important primary clay deposits are considered as china clays. Industrially significant primary clay deposits in the United States are found in North Carolina with minor deposits in Pennsylvania, California, and Missouri [22]. Perhaps the most famous primary china clay deposits

Table 3   Typical compositions (weight percent) of some primary china clays [3,22]

Location	SiO_2	Al_2O_3	Fe_2O_3	TiO_2	CaO	MgO	K_2O	Na_2O	H_2O^a
North Carolina	46.2	38.4	0.6	Trace	0.4	0.4	0.6	0.1	13.3
California	45.3	38.6	0.3	Trace	0.1	0.2	1.0	1.4	13.3
England	48.3	37.6	0.5	0.2	0.2	Trace	1.3	0.3	12.0

^{a}Called "loss on ignition" in most older texts, now considered chemically combined water

are those found in Cornwall, England, the source of English china clay [22]. Typical compositions of some primary china clays (after removal of accessory minerals) are given in Table 3 [3,22]. In the raw state, high purity clays can be nearly white in color, although commercial deposits vary in color from white to ivory. Likewise, the color upon firing varies from white to ivory depending upon the impurity content. The highest quality clays are termed "white burning" because of the lack of coloring from impurities after heating.

2.3.2 Secondary Clays

Secondary or sedimentary clays are formed in one location and then transported to the location of the deposit by the action of wind or water. Often, mineral impurities present in the primary deposit are left behind during transport. Impurity minerals such as quartz and mica are almost completely removed in some cases. However, other impurities such as TiO_2 and Fe_2O_3 are often picked up during transport [3]. Secondary clay deposits tend to have distinct layers due to repeated cycles of active deposition and inactivity [6]. Secondary deposits can also be significantly larger than primary deposits and contain a wider variety of clay mineral types, since clay can be transported in from different primary deposits [6]. Major U.S. commercial deposits of secondary china clays are found in Georgia, Florida, and South Carolina, with additional deposits in Alabama and Tennessee. Typical compositions of secondary clays are given in Table 4 [22,23]. As with primary clays, the color of raw secondary clays varies with the impurities. Many deposits are white to ivory colored, but secondary clays can also be red or brown due to other impurities. Likewise after firing, color depends strongly on the impurities present.

2.4 Clays Used in the Ceramics Industry

In this section, clays are categorized based on how they are used in the ceramics industry. The two major types of ceramic clays are china clay and ball clay. Other materials of note include fire clays, bentonite, and talc. Less refractory materials including those classified as shales and stoneware clays are also of interest. The composition, important properties, and uses for these types of clays are discussed in this section.

2.4.1 China Clay

China clays, also referred to as kaolins, are used to produce traditional ceramics when the color of the finished object and its high temperature performance are important.

Table 4 Typical compositions (weight percent) of some secondary china clays [3,22,23]

Location	SiO_2	Al_2O_3	Fe_2O_3	TiO_2	CaO	MgO	K_2O	Na_2O	H_2O
Georgia	45.8	38.5	0.7	1.4	Trace	Trace	Trace	Trace	13.6
Florida	45.7	37.6	0.8	0.4	0.2	0.1	0.3	0.1	13.9
South Carolina	45.2	37.8	1.0	2.0	0.1	0.1	0.2	0.2	13.7

China clays generally contain large proportions of the mineral kaolinite, but can contain substantial amounts of other clay minerals. In all cases, the content of Fe_2O_3, TiO_2, and other potential coloring impurities is low, resulting in bodies that range in color from white to ivory. China clay is found in both residual and secondary deposits. As detailed in Tables 3 and 4, the compositions of most china clays are slightly Al_2O_3 poor compared with the composition expected based on the mineralogical composition of kaolinite ($Al_2O_3\ 2SiO_2\ 2H_2O$ is 46.6 wt% silica, 39.5 wt% alumina, and 13.9 wt% water) due to the presence of impurities. China clays tend to have a moderate particle size (1–2 μm). Because of the particle size, china clays produce moderate plasticity during forming compared with other clays. Drying and firing shrinkage also tend to be moderate. China clays are used in many traditional ceramics, including pottery and stoneware, along with refractories and finer ceramics such as hard porcelains.

2.4.2 Ball Clay

Ball clays are remarkable because of their high plasticity when mixed with water. The plasticity is a result of fine particle size (0.1–1 μm), which is stabilized by a substantial content of organic matter (up to 2 wt%). Typical compositional data for ball clays are given in Table 5 [24]. Because of the fine particle size, the water demand for ball clays is higher than for most china clays. The fine particle size also gives ball clay bodies higher green strength and higher fired strength than other clays. In the raw state, ball clays range in color from light brown to nearly black, depending heavily on the organic content. After firing, the higher Fe_2O_3 and TiO_2 contents give ball clays, compared to china clays, an ivory to buff color. Ball clays are used extensively in whitewares, pottery, and traditional ceramics due to the workability and strength. However, their use in hard porcelains or other applications where color is important is minimal.

2.4.3 Fire Clay

Though no standard definition exists, the term fire clay refers to secondary clays that are not ball clays or china clays, but can be used to produce refractory bodies [3]. Fire clays are often found in proximity to coal deposits, but this is not true for all fire clays or for all coal deposits [6]. The main sub-types of fire clays, in the order of increasing alumina content, are plastic fire clays, flint fire clays, and high-alumina fire clays. The compositions of typical fire clays are summarized in Table 6 [22]. Among the attributes common to the different varieties of fire clays are their relatively low concentration of fluxing impurities (alkalis, alkaline earths) and their non-white color after firing. Through the 1970s, refractories made from fire clays set the standard for performance in metal processing applications due to their low cost, high corrosion

Table 5 Typical compositions (weight percent) of some ball clays [3,24]

Location	SiO_2	Al_2O_3	Fe_2O_3	TiO_2	CaO	MgO	K_2O	Na_2O	H_2O
Tennessee	57.6	28.1	1.1	1.4	Trace	Trace	0.9	0.1	10.6
Tennessee	51.7	31.2	1.2	1.7	0.2	0.5	0.4	0.6	12.1
Kentucky	57.7	28.5	1.2	1.5	0.2	0.2	0.1	1.2	9.5

Table 6 Typical compositions (weight percent) of common types of fire clays [22]

Type	SiO_2	Al_2O_3	Fe_2O_3	TiO_2	CaO	MgO	K_2O	Na_2O	H_2O
Plastic fire clay	58.1	23.1	2.4	1.4	0.8	1.0	1.9	0.3	8.0
Flint fire clay	33.8	49.4	1.9	2.6	–	–	–	–	12.0
Diasporitic fire clay	29.2	53.3	1.9	2.7	–	–	–	–	12.0

resistance, and excellent thermal stability. However, higher processing temperatures and increasingly stringent batch chemistry requirements have driven most industries to alternative refractory linings such as high-alumina castables, basic brick, or carbon containing materials. In spite of the shift in industry needs, fire clay refractories are still used extensively. Current uses for fire clay refractories include insulation behind hot-face materials, low heat duty furnace linings, and specialty applications such as laboratory crucibles and setters.

Plastic fire clays have a composition similar to china and ball clays, except for the elevated Fe_2O_3 and TiO_2 contents. Because of their composition, plastic fire clays have similar plasticity, dried strength, and fired strength when compared with china clays. Plastic fire clays range in color from gray to red or even black in the raw state. Like other fire clays, plastic fire clays produce buff-colored bodies when fired.

Flint fire clays have a higher alumina content than plastic fire clays, ball clays, and china clays, in addition to having slightly elevated levels of Fe_2O_3 and TiO_2 (Table 6) [22]. Flint fire clays have lower plasticity (compared with china clays) when mixed with water and, consequently, develop lower dried and fired strengths. Because of the lower plasticity, the drying and firing shrinkages tend to be very low [25]. Processing of flint fire clays can require plastic additives such as ball clays or bentonites. In the raw state, flint fire clays range in color from gray to red and flint fire clay deposits tend to be harder than other clays [3].

High-alumina fire clays found in the U.S. contain substantial amounts of alumina minerals such as diaspore, in addition to the aluminosilicate clay minerals present. High-alumina fire clays can have much higher alumina content than other common clays (Table 6). These clays produce refractory bodies when fired, but have comparatively low plasticity when mixed with water. Like flint fire clays, high-alumina fire clays undergo little shrinkage when dried or fired. In addition, the dried strength of bodies produced from high-alumina fire clays is poor. High-alumina fire clays tend to be gray to reddish-brown or brown in the raw state and produce buff-colored objects when fired.

2.4.4 Bentonite

Bentonites are highly plastic secondary clays that are used in small amounts as absorbents or as binders/plasticizers in batches of other materials [3]. Bentonites are formed from volcanic ash or tuff rather than igneous rocks [6]. The most significant commercial deposits of bentonite in the U.S. are in Wyoming, but bentonite deposits are widespread. The main crystalline constituent of bentonites is montmorillonite, with Mg and Fe substitution onto the octahedral sites (Fig. 6). Bentonites swell significantly when mixed with water. Also, bentonites form highly thixotropic gels, even in low concentration [14]. Because of swelling and extremely high drying and

Table 7 Typical compositions (weight percent) of other commonly used clays

Type	SiO_2	Al_2O_3	Fe_2O_3	TiO_2	CaO	MgO	$K_2O + Na_2O$	H_2O
Bentonite	49.6	15.1	3.4	0.4	1.1	7.8	–	23.0
Talc	56.3	3.2	5.4	–	0.4	27.9	0.9	5.7
Shale	54.6	14.6	5.7	–	5.2	2.9	5.9	4.7

firing shrinkages, bentonites are rarely used as a major constituent of traditional ceramics; applications are confined to additives in a variety of processes. The typical composition of a bentonite is given in Table 7 [6].

2.4.5 Talc

Talc is the magnesium silicate structural analog to pyrophyllite. Its properties are nearly identical to pyrophyllite, except that Al^{3+} cations have been replaced by Mg^{2+} cations [25]. Talc occurs in secondary deposits and is formed by the weathering of magnesium silicate minerals such as olivine and pyroxene [2]. In bulk form, talc is also called soapstone and steatite. A typical composition for talc is given in Table 7 [22]. Historically, talc has been used extensively in electrical insulator applications, in paints, and as talcum powder [2].

2.4.6 Shales

Shale is a term that refers to sedimentary deposits that have been altered by compaction and, in some cases, the cementation of grains by deposition of other minerals such as sericite (a fine grained muscovite) [6]. Shales are identical structurally and chemically to clays, although the water content of shales tends to be significantly lower. However, when they are mixed with water, shales develop plasticity similar to clays and can be used interchangeably [22]. In fact, weathering of shales is one method for the formation of clays [2]. Shales often contain high levels of iron, giving them a red color when fired [22]. A typical shale composition is given in Table 7 [22].

2.4.7 Other Clays

An enormous variety of other grades of clay minerals have been used commercially. The names of these clays can be based on the ultimate application (stoneware clay, brick clay) or the fired properties (red firing clay, vitrifying clay) [25]. The compositions of these clays are highly variable, but in general they contain high amounts of alkalis and high amounts of Fe_2O_3, TiO_2, and other impurities. A few of these clays (earthenware, stoneware, and brick clays) are mentioned here, but interested readers should consult references [3,6,14,22, 25] for information on other categorizations as well as chemical composition information. Keep in mind that the designations are not based on mineralogy, composition, or any specific property. A particular clay may fall under one or more categories depending on how it is gathered, its beneficiation, or its intended use.

Earthenware refers to products produced from unbeneficiated clays with no other additives. Earthenware clays are formed by incomplete conversion of the parent mineral formation and they contain substantial amounts of residual feldspar and quartz, giving a composition similar to a triaxial whiteware [3]. Earthenware bodies are typically formed by throwing or modeling [22]. Earthenwares are self-fluxing during firing due to the alkali content. Fired earthenware bodies typically have high absorption (10–15%) and are fired at moderate temperatures (cone 5–6) [22]. Fired earthenware bodies are usually red and find use as decorative objects, as tiles, or as tableware [26].

Stoneware clays can be used without beneficiation or additives to produce ware with low absorption (0–5%) at relatively low temperatures (cone 8–9) [22]. Fired stoneware objects usually have a buff or gray color and are used as electrical insulators, cookware, decorative items, drain pipe, tiles, and tableware [26]. Stoneware can be formed by casting, throwing, or pressing. The major difference between stoneware clays and earthenware clays is Fe_2O_3 content, with stoneware clays usually having lower Fe_2O_3 than earthenware.

Brick clays tend to be high in alkalis and iron, but low in alumina [14]. The clays usually have moderate to high plasticity, which facilitates forming [25]. Often, brick clays are actually shales [14]. These clays fire at moderate temperatures (cone 1–5) and the resulting fired bodies are dark red. Clays with similar properties but different colors upon firing can be used to produce other products such as sewer tile and roofing tile [6]. Nearly any red burning clay can be classified as brick clay.

3 Processing Methods for Clay-Based Ceramics

The study of clay-based ceramics has an enduring legacy due to the science that developed to understand the rheological behavior of clay–water pastes. As stated repeatedly in this chapter, clays develop plasticity when mixed with water. Plasticity, as defined by Grim, is "the property of a material which permits it to be deformed under stress without rupturing and to retain the shape produced after the stress is removed" [14]. For countless generations, clay-based ceramics were formed by mixing clay and other ingredients with some amount of water (determined by trial and error and/or experience) to get a consistency (i.e., rheology) that was acceptable for the forming method of choice. As new analytical tools were developed throughout the twentieth century, ceramists used them to examine the structure of clay minerals and to understand how clays interacted with water. Even though the emphasis in the field of ceramic engineering has shifted away from traditional ceramics to advanced materials, processing science still focuses on processing methods (dry pressing, extrusion, tape casting, and slip casting) that rely on controlled plastic deformation during forming, thus mimicking the behavior of clay–water pastes [1]. The key difference is that advanced materials use organic additives to promote plasticity whereas plasticity develops naturally when water is added to clays.

3.1 Clay–Water Interactions

The processing methods for clay-based ceramics can be categorized by the water content and the resulting rheological behavior. The methods that will be discussed in this section, in order of increasing water content, are: (1) dry pressing (2) stiff plastic forming, (3) soft plastic forming, and (4) casting. Most clay compositions can be fabricated using any of the forming processes by simply changing the water content of the batch. As such, the choice of forming methods is often dictated by the desired shape of the product and will be discussed in that context. Water contents and shape limitations for the four forming methods are summarized in Table 8. The overlap in the water contents for the different techniques is due to the varying water requirements for different clays, which is caused by differences in composition, structure, and physical characteristics of the clays.

Forming techniques used for clay-based ceramics require control of water content in the batch. Water content, in turn, affects the response of the clay during forming [27]. As the water content of the batch increases, the yield point of the clay–water mixture, and thus the force required to form the desired shape, generally decreases [26]. However, the relationship is complex and depends on the composition of the clay, its structure, additives to the batch, and other factors [14]. One method for quantifying the behavior of clay–water pastes is to measure the plastic yield point as a function of water content [14]. The water contents and maximum yield points in torsion are compared for several clays in Table 9. Kaolins and plastic fire clays require the least amount of water to develop their maximum plasticity, ball clays require an intermediate amount, and bentonite requires the most.

The interactions between water and ceramic particles are complex and important for processes ranging from the rheology of slurries to the drying of particulate solids. An in-depth discussion of water–particle interactions is beyond the scope of this chapter. For the discussions that follow, it is sufficient to understand the forms that water takes within a particulate ceramic [27]. At the lowest contents, water is present as partial, complete, or multiple layers adsorbed (physical) on the surface of the particles. After the surfaces are covered with a continuous adsorbed film, liquid water can condense in the pores between particles. Finally, at the highest water

Table 8 Water contents and pressure range used during the four common forming methods used for clay-based ceramics [22,26]

Method	Water (wt%)	Pressure range (MPa)
Dry pressing	0–15	100–400
Stiff plastic	12–20	3–50
Soft plastic	20–30	0.1–0.75
Slip casting	25–35	None

Table 9 Water content and maximum yield point for different types of clays [14]

Clay	Water content (wt%)[a]	Yield torque (g cm^{-1})
Kaolin	19.2	472
Plastic fire clay	19.0	442
Ball clay	34.4	358
Bentonite	41.9	254

[a]Determined as weight of water added to clay dried at 105°C for 24h

contents, free water that does not interact with particle surfaces begins to separate individual particles, eventually leading to a stable dispersion of fine, separated particles.

3.2 Dry Pressing

Dry pressing refers to forming methods that require up to 15 wt% water in which plastic deformation of the clay–water mixture is minimal. At the lower end of the water contents, water is present as a partially adsorbed layer. At the higher end, the particle surfaces will be completely covered by the adsorbed layer and some water will condense in fine pores. The amount of water needed for a pressing operation varies depending on the pressing characteristics desired, the state of hydration of the clay, how the clay interacts with water, and the particle size of the clay [1,22]. In dry pressing, water acts mainly as a binder that promotes green strength in a compacted body.

Dry pressing is defined as the simultaneous shaping and compaction of a powder in either a rigid die or a flexible container [28]. Common variations on the technique include uniaxial pressing and isostatic pressing [29]. The water content must be sufficient to promote binding of the clay particles without forming a continuous water film that would allow for excessive plastic deformation under an applied load. Dry pressing is the most common forming technique used in the ceramics industry and it is used to form a variety of clay-based ceramics including floor and wall tile, bricks, and electrical insulators [29]. Shapes with a low aspect ratio (height to diameter) are commonly formed by pressing operations [29]. A schematic representation of a die used for uniaxial dry pressing, along with the resulting forces on the powder compact, is shown in Fig. 8. Compaction pressures range from 20 to 400 MPa (3–60 ksi) with an upper pressure limit of around 100 MPa for uniaxial pressing. Fabrication of parts with high aspect ratios or the use of pressing pressures above 100 MPa can lead to the development of pressure gradients (Fig. 9) and other defects that affect the quality of parts after pressing and after firing [29]. As a side note, most nonclay ceramics require the addition of binders and plasticizers as

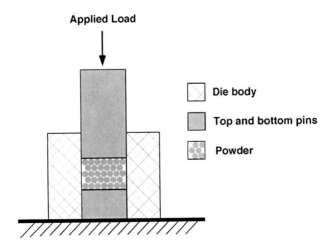

Fig. 8 Schematic representation of a common die geometry for dry pressing [1]

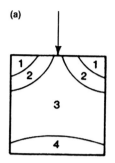 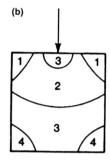

Fig. 9 Schematic representation of pressure gradients that are present in uniaxially pressed clay-based ceramics after pressing at (**a**) low and (**b**) high pressure (Reproduced by permission of John Wiley from J.S. Reed, *Principles of Ceramic Processing*, 2nd Edition, John Wiley, New York, 1995) [1]

forming aids [1]. Organic additives are commonly used as binders and plasticizers, but clays such as bentonites are also used as binders/plasticizers in many applications [14].

3.3 Stiff Plastic Forming

The water content for stiff plastic forming techniques is between 12 and 20 wt%, which produces partial or full filling of pores by water [1,25]. Extrusion is the most common stiff plastic forming technique, although injection molding can also fall under this category [22]. The pressures required for stiff plastic forming are lower than dry pressing, ranging from ~3 to 50 MPa (~0.5–10 ksi) due to the higher water content, which results in lower plastic yield points [26]. Extrusion is used to form clay-based products with a uniform cross section such as pipe, tubes, rods, and bricks [1]. In addition, thin-walled products with fine structure details such as honeycomb supports for catalytic converters can be extruded [1]. Extrusion processes can either be continuous or batch type [30]. Continuous auger extruders mix raw materials in a pug mill, shred and de-air the resulting plastic mass, and then force it through a die (Fig. 10a) [30]. The shape of the die opening and the positioning of "spiders" or other tooling in the throat of the die determine the shape of the extruded part [1]. The piston extruders used for batch processes can be used to form the same shapes as continuous auger extruders and they have a much simpler design (Fig. 10b). However, piston extruders can only produce limited quantities of product from a premixed plastic mass [30]. Common defects in extruded parts include laminations caused by wall friction and crow's foot cracks around rigid inclusions [1]. Nonclay ceramics can also be formed by extrusion, but require formulation of suitable binder/plasticizer combinations [1].

3.4 Soft Plastic Forming

The water content for soft plastic forming methods ranges from 20 to 30 wt%, which produces complete filling of pores by water and can result in some additional free water that separates the particles in the structure [22,31]. Soft plastic forming techniques include high-volume mechanical techniques such as jiggering, jollying, and ram pressing

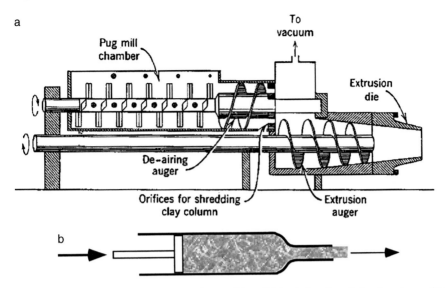

Fig. 10 Augur (**a**) and piston (**b**) type extruders used for stiff plastic forming of clay-based ceramics. (**a** reproduced by permission of John Wiley from W.D. Kingery, *Introduction to Ceramics*, 1st Edition, John Wiley, New York, 1960; **b** reproduced by permission of the McGraw-Hill Companies from F.H. Norton, *Fine Ceramics*, McGraw Hill, New York, 1970) [25,3]

and low-volume hand techniques such as throwing and wheel turning [27]. The imposed pressure range for soft plastic forming is 0.1–0.75 MPa (15–110 psi) [26]. The soft plastic forming techniques of jiggering and jollying are used to form objects that have a center of radial symmetry conducive to forming by rotation, such as dinner plates, cups, mugs, and flowerpots. During jiggering, the plastic mass is placed on a rotating form that determines the profile of the product and a tool is brought down to cut away excess material from the back (Fig. 11a) [31]. Jollying is similar to jiggering, except that in this case the tool determines the inner profile of the product and the form serves as a physical support [26]. Typically, objects such as dinner plates are formed by jiggering, while objects such as teacups and mugs are formed by jollying. Similarly, ram pressing employs two forms that are pressed together, but without rotating tooling [26]. Ram pressing can be used to form a variety of shapes including dinner plates (Fig. 11b).

3.5 Casting

Slip casting is used to produce clay-based ceramics from clay–water slurries containing 25 wt% water or more [26]. For casting slips, the water content is high enough so that all of the particles in the system are separated by free water. Most often, slip casting requires no applied pressure, although many industrial shops have switched to pressure casting (slip casting with an applied pressure) to improve productivity and reproducibility. Slip casting requires a well-dispersed, stable suspension of ceramic particles and a porous mold, which is most often gypsum (hydrated plaster of Paris) [22]. When the slurry is poured into the mold, the pores in the mold draw water out of the slurry, causing particles to deposit on the mold surface (Fig. 12) [15]. When the cast layer has sufficient thickness, the excess slip is poured out, leaving a thin, negative replica of the mold. The replica is partially dried in the mold until it pulls away

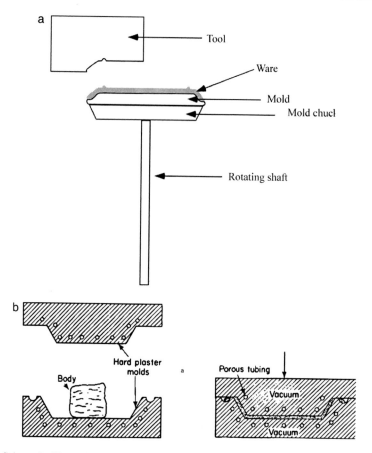

Fig. 11 Schematic illustrations of the processes of (**a**) jiggering and (**b**) ram pressing used for soft plastic forming of clay-based ceramics (**a** reprinted by permission of Addison-Wesley from F.H. Norton, *Elements of Ceramics*, Addison-Wesley Publishing, Reading, MA, 1952; **b** reprinted by permission of the McGraw-Hill Companies from F.H. Norton, *Fine Ceramics*, McGraw Hill, New York, 1970) [22,3]

from the sides of the mold and is rigid enough to be removed without deformation. Slip casting is used to form objects that are not conducive to jiggering or jollying because of a lack of radial symmetry or those with complex surface details such as decorative figurines. Slip casting is also used to form objects that are difficult to form by other techniques because of their size such as radomes, large crucibles, and large diameter furnace tubes [25]. Slip casting does not work as well with solid objects due to the problem of removing water uniformly and differential shrinkage [1].

4 Kaolinite to Mullite Reaction Sequence

When heated, kaolinite undergoes a complex series of chemical and physical changes that transform the layered mineral to a combination of crystalline mullite and an amorphous siliceous phase. Though simple conceptually, the study of this reaction sequence

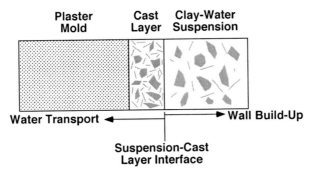

Fig. 12 Schematic representation of the build up of a cast layer (wall) from a suspension of clay particles (slip) in contact with a gypsum(hydrated plaster of Paris) mold (reproduced by permission of Addison-Wesley from F.H. Norton, *Elements of Ceramics*, Addison-Wesley Publishing, Reading, MA, 1952) [22]

continues to draw interest from the materials research community due to on-going controversies related to the composition and structure of the intermediate phases. Notable studies of the reaction of clays during heating have been conducted by LeChatelier [32], Brindley [33], and MacKenzie [34]. In an interesting parallel to the connection between the processing of clay-based ceramics and advanced processing methods, the characterization protocols used in modern ceramic science draw heavily on the work of these authors who were among the first in the field of materials to apply characterization techniques that are now considered routine. LeChatelier used thermal analysis, Brindley employed a combination of transmission electron microscopy and diffraction, and MacKenzie made use of nuclear magnetic resonance spectroscopy.

4.1 Loss of Adsorbed Water

At temperatures below 150°C, water that is physically attached to clays evaporates. Physically attached water can be present as water adsorbed onto the surface of particles or between the layers of the clay structure. The loss of water is endothermic and results in measurable weight loss. For kaolinite, the weight loss is usually minor, on the order of a percent or less. However, other clays, particularly those that swell when exposed to water such as bentonite can have considerable weight loss in this temperature regime (4–8 wt%) [6]. For kaolinite, the changes due to loss of physical water do not alter the structure as determined by X-ray diffraction.

4.2 Metakaolin

Around 450°C, the chemically combined water in clays is released, resulting in the formation of metakaolin. As with the loss of physically adsorbed water, the loss of the chemical water is an endothermic process that is accompanied by weight loss [33]. The magnitude of the weight loss depends on the amount of chemically combined water in the clay. For kaolinite, $Al_2O_3 \cdot 2SiO_2 \cdot 2H_2O$, the weight loss due to chemically combined water should be 13.9 wt%, which is similar to reported water contents for high-purity secondary kaolins (Table 4). After dehydration, metakaolin appears amorphous on X-ray diffraction, but the short range ordering of the cations within the sheets that make up the kaolinite structure is retained [33,34]. Brindley has speculated that disruption of the order perpendicular to the sheets causes the change in the X-ray diffraction pattern [33]. Thus, metakaolin is a homogeneous molecular-level mixture of noncrystalline alumina and silica. Metakaolin does not spontaneously rehydrate when it is exposed to water and it remains stable up to approximately 980°C.

4.3 Spinel

As metakaolin is heated, it undergoes a structural transformation around 980°C, a temperature of significant interest in the synthesis of mullite ceramics [35]. Brindley, among others, has observed the formation of spinel, an amorphous siliceous phase, and a small amount of nanocrystalline mullite at 980°C [33]. This process is exothermic with no accompanying weight loss. The observed heat of reaction comes mainly from spinel formation [33]. Investigators are in general agreement that the spinel phase is similar in structure to the cubic transitional alumina γ-Al_2O_3 and that it contains most of the alumina from the original kaolin. The amorphous phase is mainly silica, but it also contains a small amount of alumina plus most of the impurities from the original clay. The mullite phase makes up only a small volume fraction of the total volume after heating to 980°C and is composed of submicrometer needle-like mullite grains. Questions remain regarding the composition of the spinel phase, with proposed compositions ranging from pure γ-Al_2O_3 to $2Al_2O_3 \cdot 3SiO_2$, which includes the mullite composition, $3Al_2O_3 \cdot 2SiO_2$ [33,36]. It seems unlikely that spinel is pure γ-Al_2O_3, since mixtures of γ-Al_2O_3 and silica prepared from colloidal particles form α-Al_2O_3 and amorphous silica around 1200°C prior to mullite formation at higher temperatures [37]. It also seems unlikely that spinel is poorly crystalline mullite, at least after heating to 980°C, since a second mullite crystallization event is recorded at higher temperatures [37]. Recent studies using nuclear magnetic resonance spectroscopy indicate that the spinel phase formed at 980°C may contain just a few weight percent silica [38]. Logically, the composition of the spinel phase probably lies between that of metakaolin ($Al_2O_3 \cdot 2SiO_2$) and mullite ($3Al_2O_3 \cdot 2SiO_2$) and is part of a phase separation process that leads to the eventual formation of mullite and an amorphous silica-rich phase [33].

4.4 Mullitization

As kaolinite is heated beyond 980°C, the small fraction of mullite crystals that formed at 980°C continue to grow, albeit at a slow rate. Mullite growth is accompanied by the disappearance of the spinel phase, although the amount of mullite formed is lower than expected based on the spinel loss [33]. Mullite formation does not approach completion until a second exothermic event occurs at approximately 1200°C, as recorded by differential thermal analysis [33]. When formed by solid-state reaction, mullite has a composition of $3Al_2O_3 \cdot 2SiO_2$, approximately 72 wt% alumina and 28 wt% silica [39]. According to Brindley, the mullite formed by heating to 1200°C contains all of the alumina from the original clay, while the silica is distributed between the mullite phase and an amorphous phase [33]. Further heating alters the size of the needle-like mullite grains and can result in crystallization of the silica to crystobalite [22]. Heating to 1200°C is generally sufficient to fully densify clay-based ceramic bodies.

5   General Outlook

Clays will continue to be an important industrial mineral for the foreseeable future. Clays continue to be used widely as raw materials for refractories and other traditional ceramics because of their availability, low cost, and ease of processing. However, a majority of applications for clay minerals lie outside the field of ceramics, as summarized in Tables 10 and 11 and described in detail in several of the references [8,14,21]. Because of this breadth of applications and continued availability of easily-mined, high-quality clay deposits, the current level of production and utilization of clay minerals should continue [8]. Production is currently stabilized around 40 million metric tons per year with an average price of approximately $30 per ton [8]. More importantly for the modern materials community, understanding the processing and characterization of traditional ceramics can provide significant insight into the structure of the materials curriculum and the methods used to process and characterize advanced ceramic materials.

Table 10   Important applications for clay minerals grouped by clay type and application [8]

Clay	Application	Important Property(ies)
Kaolin	Paper	Absorbency, color
	Refractories	High temperature stability
	Traditional ceramics	Plasticity, fired strength
Ball Clay	Traditional ceramics	Plasticity, fired strength
Bentonite	Kitty litter	Absorbency
	Foundry sand	Binding ability
	Iron ore palletizing	Binding ability
	Petroleum drilling mud	Viscosity control, thixotropy
Fire Clay	Refractories	High temperature stability

Table 11   Other applications for clay minerals

Application	Reason for use
Absorbent	Water affinity
Adhesive	Viscosity control, inert filler
Aggregate	Low cost, low density
Cement	Al_2O_3 source
Clarification of beverages	Surface charge characteristics
Paint	Pigment, inert filler, viscosity control
Paper	Brightness, absorbency
Petroleum refining	Catalytic activity

Acknowledgments Many of the references used in the preparation of this chapter were donated to the Materials Science and Engineering department at MS&T from the A.P. Green Industries, Inc. corporate library. The author would like to thank Dr. David J. Wronkiewicz of the MS&T Geology and Geophysics program, Dr. Jeffrey D. Smith of the MS&T Materials Science and Engineering Department, and Ms. Jill C. Fahrenholtz for their critical reviews of various drafts of the manuscript.

References

1. J.S. Reed, *Principles of Ceramic Processing*, 2nd edn., John Wiley, New York, 1995.
2. C.S. Hurlbut, *Dana's Manual of Mineralogy*, 15th edn., John Wiley, New York, 1941.
3. F.H. Norton, *Fine Ceramics*, McGraw Hill, New York, 1970.
4. L. Jiazhi, The evolution of chinese pottery and porcelain technology, in *Ancient Technology to Modern Science, Ceramics and Civilization,* Vol. 1, W.D. Kingery (ed.), The American Ceramic Society, Columbus, OH, 1984, pp. 135–162.
5. L. Gouzhen and Z. Xiqiu, The development of chinese white porcelain, in *Ancient Technology to Modern Science, Ceramics and Civilization,* Vol. 2, W.D. Kingery (ed.), The American Ceramic Society, Columbus, OH, 1985, pp. 217–236.
6. H. Ries, *Clays: Their Occurrence, Properties, and Uses*, 3rd edn., John Wiley, New York, 1927.
7. F.H. Norton, Cycles in ceramic history, in *Ceramic Processing Before Firing*, G.Y. Onoda and L.L. Hench (eds.), John Wiley, New York, 1978, pp. 3–10.
8. Mineral Commodity Survey 2003, U.S. Geological Survey, 2003.
9. 2001 Annual Survey of Manufacturers, U.S. Census Bureau, 2002.
10. T.L. Brown and H.E. LeMay, Jr., *Chemistry: The Central Science*, 2nd edn., Prentice-Hall, Englewood Cliffs, NJ, 1981, pp. 678–685.
11. C. Klein and C.S. Hurlbut, *Manual of Mineralogy*, 21st edn., John Wiley, New York, 1993.
12. W.D. Neese, *Introduction to Optical Mineralogy*, Oxford University Press, New York, 1986, pp. 234–251.
13. G.W. Brindley, Ion exchange in clay minerals, in *Ceramic Fabrication Processes*, W.D. Kingery (ed.), John Wiley, New York, 1958, pp. 7–23.
14. R.E. Grim, *Applied Clay Mineralogy*, McGraw-Hill, New York, 1962.
15. W.D. Kingery, H.K. Bowen, and D.R. Uhlmann, *Introduction to Ceramics*, 2nd edn., John Wiley, 1976, pp. 77–80.
16. M.C. Gastuche, The octahedral layer, in *Clays and Clay Minerals, Proceedings of the Twelfth National Conference on Clays and Clay Minerals*, W.F. Bradley (ed.), Pergamon Press, Oxford, 1963, pp. 471–493.
17. B. Velde, *Clay Minerals: A Physico-Chemical Explanation of their Occurrence*, Elsevier, Amsterdam, 1985, pp. 1–44.
18. G.W. Brindley, Structural mineralogy of clays, in *Clays and Clay Technology, Proceedings of the First National Conference on Clays and Clay Technology*, J.A. Pask and M.D. Turner (eds.), California Division of Mines, San Francisco, 1955, pp. 33–44.
19. P.F. Kerr, Formation and occurrence of clay minerals, in *Clays and Clay Technology, Proceedings of the First National Conference on Clays and Clay Technology*, J.A. Pask and M.D. Turner (eds.), California Division of Mines, San Francisco, 1955, pp. 19–32.
20. W. Ryan, *Properties of Ceramic Raw Materials*, 2nd edn., Pergamon Press, Oxford, 1978, pp. 42–72.
21. *Kaolin Clays and Their Uses*, 2nd edn., The Huber Chemical Corporation, New York, 1955.
22. F.H. Norton, *Elements of Ceramics*, Addison-Wesley, Reading, MA, 1952, pp. 1–35.
23. EPK Kaolin Product Information Sheet, Zemex Industrial Minerals, Atlanta, GA, 2002.
24. S-4 and 44-B Clay Product Information Sheets, Old Hickory Clay Company, Gleason, TN, 2001.
25. W.D. Kingery, *Introduction to Ceramics*, 1st edn., John Wiley, New York, 1960.
26. R.A. Haber and P.A. Smith, Overview of traditional ceramics, in *Ceramics and Glasses: Engineered Materials Handbook*, Vol. 4, S.J. Schneider, Jr. (ed.), ASM International, Materials Park, OH, 1991, pp. 3–15.

27. F.H. Norton, Clay–Water Pastes, in *Ceramic Fabrication Processes*, W.D. Kingery (ed.), John Wiley, New York, 1958, pp. 81–89.

28. S.J. Glass and K.G. Ewsuk, Ceramic powder compaction, *MRS Bulletin*, 1997, pp. 24–28.

29. B.J. McEntire, Dry pressing, in *Ceramics and Glasses: Engineered Materials Handbook*, Vol. 4, S.J. Schneider, Jr. (ed.), ASM International, Materials Park, OH, 1991, pp. 141–146.

30. I. Ruppel, Extrusion, in *Ceramics and Glasses: Engineered Materials Handbook*, Vol. 4, S.J. Schneider, Jr. (ed.), ASM International, Materials Park, OH, 1991, pp. 165–172.

31. R.E. Gould and J. Lux, Some experiences with the control of plastic bodies for automatic jiggering, in *Ceramic Fabrication Processes*, W.D. Kingery (ed.), John Wiley, New York, 1958, pp. 98–107.

32. M.H. LeChatelier, De L'Action De La Chaleur Sur Les Agriles, *Societe Française Minéralogie Bulletin*, **10**(6), 204–211 (1887).

33. G.W. Brindley and M. Nakahira, The kaolinite-mullite reaction series: I, a Survey of outstanding problems; II, metakaolin; and III, the high temperature phases, *J. Am. Ceram. Soc.* **42**(7), 311–324 (1959).

34. K.J.D. MacKenzie, I.W.M. Brown, R.H. Meinhold, and M.E. Bowden, Outstanding problems in the kaolinite–mullite reaction sequence investigated by silicon-29 and aluminum-27 solid-state nuclear magnetic resonance: I, metakaolinite, *J. Am. Ceram. Soc.* **68**(6), 293–297 (1985).

35. J.A. Pask, X.W. Zhang, A.P. Tomsia, and B.E. Yoldas, Effect of sol–gel mixing on mullite microstructure and phase equilibria in the α-Al_2O_3–SiO_2 system, *J. Am. Ceram. Soc.* **70**(10), 704–707 (1987).

36. I.W.M. Brown, K.J.D. MacKenzie, M.E. Bowden, and R.H. Meinhold, Outstanding problems in the kaolinite–mullite reaction sequence investigated by ^{29}Si and ^{27}Al solid-state nuclear magnetic resonance: II, high temperature transformations of meta-kaolin, *J. Am. Ceram. Soc.* **68**(6), 298–301 (1985).

37. A.K. Chakraborty and D.K. Ghosh, Reexamination of the kaolinite-to-mullite reaction series," *J. Am. Ceram. Soc.* **61**(3–4), 169–173 (1978).

38. K.J.D. MacKenzie, J.S. Hartman, and K. Okada, MAS NMR evidence for the presence of silicon in the alumina spinel from thermally transformed kaolinite, *J. Am. Ceram. Soc.* **79**(11), 2980–2982 (1996).

39. K. Okada and N. Otsuka, Formation process of mullite, in *Mullite and Mullite Matrix Composites: Ceramic Transactions*, Vol. 6, S. Somiya, R.F. Davis, and J.A. Pask (eds.), The American Ceramic Society, Westerville, OH, 1990, pp. 375–388.

Chapter 8
Concrete and Cement

Mariano Velez

Abstract Inorganic concretes are reviewed, emphasizing two major areas: construction concretes and high temperature (refractory) concretes. Although such materials are intended for completely different applications and markets, they have in common that they are made from inorganic ceramic oxides and both materials are used for structural purposes. Current applications and research topics representing new challenges are summarized.

1 Introduction

Concrete usually indicates a construction material made from Portland cement, aggregates (for instance, gravel and sand), water, and additives to improve mixing or specific properties of the final material. Refractory concretes (monolithic refractories) refer to high-temperature materials for the manufacture of shaped refractories. Most concrete refractories are based on calcium aluminates, although some applications require the use of other high-temperature ceramic materials, such as magnesium oxide. The distinguishable feature of these concretes is that the method of preparation does not involve forming or firing at the manufacturing plant as in the fabrication of refractory bricks. This compilation includes only concretes for construction and refractory applications.

Cement is a binder that sets and hardens by itself or binds other materials together. The most widely known application of cements is in construction; a second one is the area of "bone cements." Cements used in construction are characterized as hydraulic or nonhydraulic and mostly for the production of mortars and concrete. Hydraulic cements set and harden after combining with water. Most construction cements are hydraulic and based on Portland cement, which consists of calcium silicates (at least 2/3 by weight). Nonhydraulic cements include the use of nonhydraulic materials such as lime and gypsum plasters. Bone cements and bone cement composites refer to compounds that have a polymer matrix with a dispersed phase of particles. For instance, polymethylmethacrylate (PMMA) is reinforced with barium sulphate crystals (for radio-opacity) or with hydroxyapatite

J.F. Shackelford and R.H. Doremus (eds.), *Ceramic and Glass Materials:*
Structure, Properties and Processing.
© Springer 2008

to form a bioactive cement. These composites are currently used in orthopaedics for bone trauma repair.

2 Construction Concrete

Natural cementitious materials have existed for very long time; however, synthetic materials were perhaps first used by Egyptians and Chinese thousands of years ago [1, 2], and the Romans used pozzolanic materials (a volcanic rock in powder form and used to make hydraulic cement) to build the Rome Coliseum. It was only until around 1824 when Joseph Aspdin, bricklayer of England, invented Portland cement by burning finely ground chalk with finely divided clay until carbon dioxide was driven off. The sintered product was ground and named as Portland cement after the building stones quarried at Portland, England.

Portland type cements are calcium silicates, which when mixed with water form a paste, producing a hardened mass of valuable engineering properties. The clinker (the fused or partly fused by-product that after grinding becomes cement) chemistry is (by weight) 50–70% 3CS (*alite*, $3CaO \cdot SiO_2$), 15–30% 2CS (*belite*, $2CaO \cdot SiO_2$), 5–10% 3CA ($3CaO \cdot Al_2O_3$), and 5–15% 4CAF (*ferrite*, $4CaO \cdot Al_2O_3 \cdot Fe_2O_3$). Concrete is a mix of cement paste and aggregates (inert granular materials such as sand, gravel, recycled concrete), which the cement binds together into a rock-like composite. Main applications are for civil infrastructures such as buildings, highways, underground mass transit systems, wastewater treatment facilities, and marine structures. A compilation on materials, properties, working operations, and repair is given by Dobrowolski [3].

Approximately, 1.6 billion tons of Portland cements are produced worldwide annually with an estimated 5% generation of the CO_2 emission. Global cement and concrete additives (fiber and chemical additives) demand is forecast to grow 6.3% annually through 2006, driven by construction and by higher standards for concrete that require more additives per ton [4].

The US is the third largest cement producing country in the world (after China and India). The US concrete industry is the largest manufacturing sector in the United States (cement, ready-mixed concrete, concrete pipe, concrete block, precast and prestressed concrete, and related products), with over two million jobs related directly and including materials suppliers, designers, constructors, and repair and maintenance. The value of shipments of cement and concrete production exceeds $42 billion annually. Currently, there are a total of 115 cement manufacturing plants in the US, with about 75% of the total plants owned by only ten large companies: Lafarge North America, Inc., Holcim (US) Inc, CEMEX, SA de CV, Lehigh Cement Co., Ash Grove Cement Co., Essroc Cement Corp., Lone Star Industries Inc., RC Cement Cp., Texas Industries Inc. (TXI), and California Portland Cement Company.

Two types of manufacturing processes have become prevalent in the cement industry: "wet process" and "dry process." Although these processes are similar in many respects, in the older "wet process," ground raw materials are mixed with water to form a thick liquid slurry, while in the "dry process," crushed limestone is used and raw materials are mixed together, with consequent higher energy efficiency as drying is eliminated. Figure 1 represents schematically the cement manufacturing process and the rotary kiln for clinker production.

2.1   Fiber-Reinforced Concrete

Fiber-reinforced concrete (FRC) has a randomly oriented distribution of fine fibers in a concrete mix [5]. Fiber size varies with typical length less than 50 mm and diameter less than 1 mm. Many fibers have been used for concrete reinforcement (Table 1), with additions of 0.1–3% by volume of fibers, which improves the strength of the concrete by up to 25% and increasing the toughness by a factor of 4. This means that FRC is less susceptible to cracking than ordinary concrete showing longer service life.

Structural concrete is reinforced with steel to carry tensile forces that are internally generated when a structure undergoes elastic and inelastic deformations. Steel fiber-reinforced concrete (SFRC) consists of cement containing aggregates and discontinuous steel fibers (low-carbon steels or stainless steel; ASTM A-820 [6] provides a classification for steel fibers). In tension, SFRC fails after the steel fiber breaks or is pulled out of the cement matrix. The strain and force interaction is complex and depends on factors such as chemical and mechanical bond between concrete and reinforcement, time-dependent properties (creep, shrinkage), environmental aspects (freezing, chemicals), geometric configuration, location and distribution, and concrete/reinforcement volume ratio.

The applications of SFRC take advantage of the static and dynamic tensile strength, energy absorbing characteristics, toughness, and fatigue endurance of the composite [7]. Uniform fiber dispersion provides isotropic strength properties. The applications include

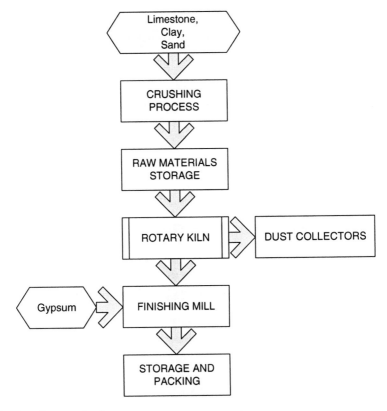

Fig. 1 Flow diagram of main process

Table 1 Fibers for concrete reinforcement

	Advantages	Disadvantages	Comparison of properties of selected materials [8]	
			Density $(g\ cm^{-3})$	Unidirectional tensile strength (GPa)
Steel	Provide very good, reasonably priced reinforcement	Corrode over time, after 6–8 years provide little reinforcement	8.0	207 (steel 4130)
Stainless Steel	Very good reinforcement	Very expensive		
Glass fiber	Good reinforcement	Alkaline nature of concrete causes the strength of silica-based fibers to degrade with time	1.99 (E-glass, S-glass)	52–59
Carbon and Kevlar	Excellent reinforcement, high strength	Very expensive; brittle behavior (Fig. 2)	1.55–1.63 (Carbon)	145–207
Plastic fiber	Good reinforcement	Low cost	1.38 (reinforced epoxy aramid)	83

cast-in-place SFRC (slabs, pavements industrial floors), precast SFRC (vaults and safes for instance with fiber content from 1 to 3 vol%), shotcrete (a sprayed concrete developed for civil construction, for instance in slope stabilization and in repair and reinforcing of structures), and slurry infiltrated fiber concrete (called SIFCON, where a formwork mold is randomly filled with steel fibers and then infiltrated with a cement slurry, containing a much larger fiber fraction between 8–12% by volume). Corrosion of steel reinforcement and the tendency of concrete to lose bond, and reducing structural performance over time, promote the development of economical, thermodynamically stable metallic and nonmetallic, corrosion-resistant reinforcements. Therefore, other reinforcement has been developed such as polypropylene fibers (most common in the market), glass fibers, and carbon fibers.

2.2 Carbon and Organic-Based Fibers

The evolution of fiber-reinforced plastic (FRP) as reinforcement in concrete began in the 1960s to solve the corrosion problem associated with steel-reinforced concrete in highway bridges and structures [8]. This is a class of materials defined as a polymer matrix, whether thermosetting (e.g., polyester, vinyl ester, epoxy, phenolic) or thermoplastic (e.g., nylon, PET) reinforced by fibers (e.g., aramid, carbon, glass). Each of the fibers considered suitable for use in structural engineering has specific elongation and stress–strain behavior. Composite reinforcing bars have more recently been used for construction of highway bridges and it appears that the largest market will be in the transportation industry. Figure 2 shows an example of repairing a small bridge by preparing a network of FRP beams and then casting a construction concrete. The engineering benefits of these fibers include the inhibition of plastic and shrinkage cracking (by increasing the tensile strain capacity of plastic concrete), reducing permeability, and providing greater impact capacity, and reinforcing shotcrete.

Fig. 2 Setting a FRP network for casting a construction concrete in the repair of a bridge (Courtesy of D. Gremmel, Hughes Brothers)

Carbon fiber-reinforced plastic (CFRP) in grid form has demonstrated potential as a reinforcing material in lightweight concrete. A high tensile capacity allows the grid to work efficiently with compressive strength in bending. Additionally, carbon fiber may act as thermal insulator. It allows a reinforced mold for production when used in grid form. However, a CFRP grid is an expensive and brittle material, and the grid form can melt under a fire accident.

2.3 Glass Fiber-Reinforced Concrete

Glass fibers have relevance in civil engineering applications because of cost and specific strength properties. The worldwide glass reinforcement market is estimated in 2.5 million tons (2001), with an average growth of 5.4% per year [9]. Original GFRC pastes used conventional borosilicate glass fibers (E-glass) and soda-lime-silica glass fibers (A-glass), which lose strength due to high alkalinity (the pH value for concrete environment is above 12.8 [10]) of the cement-based matrix. Other compositions include S-glass (an Mg–Al–silicate of high strength), S-2 glass (an S-glass composition with surface treatment), and C-glass (a Na-borosilicate) used in corrosive environments [8]. Improved alkali-resistant fibers have compositions containing 16% zirconia. Another potential alternative is the use of novel high-alkaline-resistant Fe-phosphate glasses [11]. Extended exposure of silica-based GFRC to natural weather results in changes in mechanical properties and volumetric dimension changes. Dimensional changes in GFRC can be considerably greater than those of conventional concrete, as a result of the high cement content in the mortar matrix. Over stressing or stress concentrations can cause cracks. This can be critical in components that are overly restrained.

Table 2 Applications of GFRC [7]

General Area	Examples
Agriculture	Irrigation channels, reservoir linings
Architectural	Interior panels, exterior panels, door frames, windows
Asbestos replacement	Sheet cladding, plain roof tiles, fire-resistant pads, molded shapes and forms, pipes
Ducts and shafts	Track-side ducting for cables, internal service ducts
Fire protection	Fire doors, internal fire walls, partitions, calcium silicate insulation sheets
Buildings	Roofing systems, lintels, cellar grills, floor gratings, hollow nonstructural columns, impact-resistant industrial floors, brick façade siding panels
Housing	Single and double skin cladding, prefabricated floor and roof units
Marine	Hollow buoys, floating pontoons, marina walkways, workboats
Metal placement	Sheet piling for canal, lake or ocean revetments, covers, hoods, stair treads
Pavements	Overlays
Permanent and temporary	Bridge decking formwork, parapets, abutments, waffle forms, columns and beams
Site-applied surface bonding	Bonding of dry-blocks walls, single skin surface bonding to metal lath substrates, ultra-low-cost shelters
Small buildings	Sheds, garages, acoustic enclosures, kiosks, telephone booths
Small containers	Telecommunication boxes, storage tanks, meter encasements, utility boxes
Street furniture	Seats and benches, planters, litter bins, signs, noise barriers, bus shelters
Water applications	Low-pressure pipes, sewer linings, field drainage components, tanks

Fiber is used in larger dosages, i.e., ~2 kg m^{-3} of concrete in commercial applications. The single largest application of GFRC has been the manufacture of exterior building facade panels [12], with at least 80% of all GFRC architectural and structural components manufactured in the US Other application areas are listed in Table 2.

2.4 Testing and Evaluation

There are several techniques for evaluating civil structures [i.e., 13] with pulse echo techniques including radar (reflection and scattering of electromagnetic pulses) [14–16], impact echo (propagation, reflection, and scattering of elastic waves after mechanical impact) [17, 18] and ultrasonic pulse echo (propagation and dispersion of sound waves produced with ultrasonic transducers) [19]. The more widely known techniques are perhaps impulse response (IR) and impact-echo testing. The IR method uses a low-strain impact to send stress waves through a specimen via a sledgehammer with a built-in load cell. The maximum compressive stress at the impact point is related to the elastic properties of the hammer tip. Response to the input stress is measured with a velocity transducer [20]. The IR method is suggested for evaluation of reinforced concrete structures such as floor slabs, pavements, bridge decks and piers, fluid retaining structures, chimneystacks, and silos.

More recently, the impact-echo technique is used to determine the position of intermediate and large defects in concrete structures. It is used to assess the bonding condition between facing stones, mortar, and inner rubble core in stone masonry [21] and the structural integrity of high-performance/high-strength concrete of existing buildings using SAWS (spectral analysis of surface waves) [22]. A mechanical impact on the surface of the structure is used to generate an elastic stress wave that travels through the structure, reflects off external boundaries and internal flaws, and returns to the surface. A receiver located near the point of impact is used to measure the normal surface displacements. Fast Fourier transform (FFT) is used to determine the corresponding frequency spectrum and analyze the location of flaws or (internal or external) surfaces by using [23].

$$d = \frac{C_p}{2f}, \tag{1}$$

where C_p is the P-wave (primary/pressure wave) speed within the material ($\sim$2,000–4,000 m s^{-1} for concrete depending on age and other characteristics [24, 25]) and f is the characteristic frequency. Limitations and associated problems of the method include spectral effects of the impactor, the receiving device, and the interpretation of other reflected waves [26]. Improvements are reported using noncontacting devices for both impact generation (shock waves) and response monitoring (laser vibrometer to measure surface velocity) [27].

Flexural strength is determined according to ASTM C 947 [28] and density is determined according to ASTM C 948 [29]. GFRC made of cement, AR-glass fibers, sand, and water is a noncombustible material and should meet the criteria of ASTM E 136 [30]. Single skin GFRC panels can be designed to provide resistance to the passage of flame, but fire endurances of greater than 15 min, as defined in ASTM E 119 [31], are primarily dependent upon the insulation and fire endurance characteristics of the drywall or back-up core.

2.5 Current Research Topics

A substantial amount of research is being carried at the NSF Center for Science and Technology of Advanced Cement-Based Materials at Northwestern University [32]. A survey of the literature is kept by The American Ceramic Society [33, 34]. Four general critical areas, necessary to maintain technological leadership, have been identified by the US industry [35]: (1) *design and structural systems*, (2) *constituent materials*, (3) *concrete production, delivery, and placement*, and (4) *repair and rehabilitation* (Table 3). Topics of current scientific research include characterization of microstructural features (e.g., C–S–H gel, pore shape, heterogeneity), mechanical properties (model prediction, measurement, performance improvement), chemical behavior (e. g., in corrosive environments, hydration process, $CaCO_3$ efflorescence), characteristics of concrete at elevated temperatures. Other specific areas include test methods to determine properties (i.e., fatigue, creep, and chemical properties) of FRP, which are reproducible and reliable. Repair of structures is a relevant area that has several worldwide organizations involved. In addition to the American Concrete Institute (ACI), groups providing information include the International Concrete Repair Institute, Concrete Repair Association (U.K.), Building Research Establishment (U.K.), American Concrete Pavement Association, American Association of State Highway Transportation Officials, The Concrete Society (U.K.), The Cathodic Protection Association (U.K.), and US Army Corps of Engineers.

Table 4 summarizes recent developments regarding cement and concrete materials. Improved materials and advanced sensors are needed for very demanding applications

Table 3 Main research needs in the concrete industry [adapted from 35]

Design and Structural Systems	
Structural concrete	System survivability
Reinforced concrete	Design methodologies for reinforcement and fibrous concrete
Modeling and measurement	Interaction prediction, monitoring
High-performance concrete	Improved technologies and advanced testing methods
Technology Transfer	Accelerating technology transfer from 15 to 12 years
	Appraisal services by standard and code bodies
Fire-, blast-, and earthquake-resistant materials and systems	Smart systems for design of fire-, blast-, and heat-resistant alternative reinforced structures
	Survivability reserach
Crosscutting innovations	Concrete as part of multimaterials systems
Constituent materials	
New materials	Noncorroding steel reinforcement
	Concrete with predictable performance
	Materials with reduced shrinkage and cracking
	Reduction of alkali-silica reactions in concrete
Measurement and prediction	Prediction methods and models for permeability, cracking, durability, and performance
	Quantifying benefits of using alternative materials
Recycling	Reuse of high-alkali wastewater
	Aggregate recycling
	Reuse of cementitious materials
Concrete production, delivery, and placement	
Information and control	Intelligent, integrated knowledge systems
	Improved control over nonspecified concrete
	On-line batching control
Production	Increased applications for robotics and automation
Test methods and sensors	Improved sensing technologies and portability
	Technologies to insure performance requirements
	Nondestructive test methods
Energy and environment	Reuse and recycling issues
	Reduction of transportation energy use
	Increased use of waste from crosscutting technologies
	Life-cycle model for CO_2 impact
Repair and rehabilitation	
New repair methods	New repair materials and applications technologies
	Self-repairing concrete
Assessment tools and modeling	Nondestructive testing for stress
	Long-term monitoring of structures
Repair field	Mitigation of alkali-silica reactivity
	Corrosion-canceling technologies
	Low-maintenance, long-life repair of concrete for corrosion protection of embedded steel

such as concrete in offshore oil platforms [36] and in the nuclear industry where concrete structures have performed well [37]. However, as these structures age, degradation due to environmental effects threaten their durability. Items of note are corrosion of steel reinforcement following carbonation of the concrete or ingress of chloride ions, excessive loss of posttensioning force, leaching of concrete, and leakage of posttensioning system corrosion inhibitor through cracks in the concrete.

Table 4 Recent US patents on cement-related materials development

Assignee	Number	Short title
MBT Schweiz	6,489,032	Cement structure containing a waterproofing layer
Aso Cement	6,464,776	Dusting-inhibited cement with improved strength
Sumitomo Osaka	6,419,741	Cement clinker and cement containing the same
Sika AG	6,387,176	Polymers for high-flow and high-strength concrete
Daicel-Huels	6,376,580	Cement retarder and cement retardative sheet
Taiheiyo Cement	6,358,311	Additives for cement materials and cement materials
RoadTechs Europe	6,315,492	Road repair material comprising cement and resin
Polymer Group	6,258,159	Synthetic polymer fibers into cement mixtures
Takemoto Yushi	6,176,921	Cement dispersants
Daicel-Huels	6,114,033	Cement retarder and cement retardative sheet
U. of Michigan	6,060,163	Fiber reinforcement of cement composites
Arco Chemical	6,034,208	Copolymers as cement additives
FMC Corporation	6,022,408	Compositions for controlling alkali-silica reaction
Toho Corporation	5,997,631	Hardener for waste-containing cement articles
W. R. Grace & Co	5,938,835	Cement composition inhibit drying shrinkage
DPD, Inc.	5,935,317	Accelerated curing of cement-based materials
Lafarge Canada	5,928,420	Alkali-reactive aggregate and sulphate resistance
W. R. Grace & Co	5,840,114	Early-strength-enhancing admixture for precast cement
College of Judea	5,772,751	Light-weight insulating concrete
Elkem ASA	5,769,939	Cement-based injection grout
MBT Holding	5,728,209	Unitized cement admixture
Dipsol Chemicals	5,653,796	(Phosphorous acid) Admixture for cement
Union Oil of California	5,599,857	(Waterless) Polymer concrete composition
Dow Chemical	5,576,378	High Tg polymer for portland cement mortar
C F P I	5,567,236	Improved rheology of cement-products
Louisiana State US	5,565,028	Alkali-activated class C fly ash cement
ENCI Nederland	5,482,549	Cement and products using such cement
Takenaka Corporation	5,466,289	Ultra high-strength hydraulic cement compositions

3 Refractory Concretes (Monolithic Refractories)

Monolithic refractories, or unshaped refractories, are made from almost same raw materials as firebricks. The distinguishable feature is that the method of preparation does not involve forming or firing at the manufacturing plant. A general definition includes batches or mixtures consisting of additives and binders, prepared for direct use either in state of delivery or after adding suited liquids. The final mix may contain metallic, organic, or ceramic fibers [38, 39].

3.1 Compositions and Applications

Several ways of classifying these refractories have been listed according to physical and chemical properties or applications [40, 41]:

- Method of installation: pouring/casting, trowelling, gunning (shotcreting), vibrating, ramming, and injecting
- Use or application: materials for monolithic constructions, materials for repairs, materials for laying and forming joints (mortars)
- Type of bond: hydraulic bond with hardening and hydraulic setting at room temperature, ceramic bond with setting by sintering during firing, chemical bond (inorganic or

organic–inorganic) with setting by chemical but not hydraulic reaction at room temperature or at temperature below the ceramic bond, and organic bond by setting or hardening at room temperature or at higher temperatures

- Chemical composition: silica-based and silica-alumina-based materials, chrome, magnesia, chrome-magnesia, spinel, SiC, materials containing carbon (more than 1% carbon or graphite), and special materials (containing other oxides or materials such as zircon, zirconia, Si_3N_4, etc.)
- Bulk density: lightweight (bulk density below $1.7\,g\,cm^{-3}$) and dense castables
- Norms and Standards: for instance ISO (International Organization for Standardization), and ASTM (American Society for Testing and Materials)

Calcium aluminate cements are examples of conventional refractory castables. Development of low, ultralow cement, no-cement pumpables, and self-flow castables has increased the applications of monolithics [42]. Steel-reinforced refractories (SFRR) are used in applications that include ferrous and nonferrous metal production and processing, petroleum refining, cement rotary kilns, boilers, and incinerators. Steel fibers are added to refractory concretes to improve resistance to cracking and spalling in applications of heavy thermal cycling and thermal shock loads.

Phosphate-bonded monolithic refractories are available both as phosphate-bonded plastic refractories and phosphate-bonded castables. Phosphate-bonded plastic refractories contain phosphoric acid or an Al-phosphate solution. They are generally heat setting refractories, developing high cold strength after setting, and are highly resistant to abrasion. Phosphate-bonded castables contain no cement, and magnesia may be added as setting agent [40].

3.2   Drying and Firing of Refractory Castables

Refractory monolithic linings are dried on site by one-side heating. During drying, rapid heating rates might lead to degradation of mechanical properties, and in extreme cases, to excessive buildup of pore pressure and even explosive spalling. A slow heating rate, on the other hand, is more energy and time consuming. Drying involves coupled heat and mass transfer in a porous solid undergoing microstructural changes (i.e., pore size and shape) and chemical changes (i.e., dehydration). Steam pore pressure is the main driving force for moisture transfer, as well as the force that could cause failure of the refractory concrete when it builds above its mechanical strength. Several material properties (such as permeability, thermal conductivity, and mechanical strength) are strongly affected by temperature and moisture content during the drying process. Dewatering is affected by the coupled and interactive influences of a number of variables, which include texture, mix constitution, permeability, strength, thermal conductivity, moisture content, casting and curing practice, binder level and type, dry-out schedule, and installation geometry. A common method for improving the spalling resistance of refractory concretes has been to add organic fibers to the mixes to increase permeability.

Permeability is the material property that most influences the drying process of refractory castables [43–45]. The permeability of compressible fluids flowing through rigid and homogeneous porous media is described by the Forchheimer equation, which includes a quadratic term for the flow rate q. For small changes in pressure, the Forchheimer's equation leads to Darcy's law:

$$q = k \frac{\Delta p}{L},$$
(2)

where the pressure drop ΔP is the difference between the absolute fluid pressure at the entrance and at the exit of the sample, L is the sample thickness, and K is the coefficient of permeability, used in computer simulations [i.e., 46]. Very low permeabilities of refractory castables are measured using a vacuum decay approach [47, 48]. A vacuum decay curve is generated by monitoring the pressure change across a specimen-slab in a vacuum chamber, as a function of time.

3.3 Evaluation Techniques

3.3.1 Wall Thickness

Several methods have been proposed to nonintrusively measure the thicknesses of walls, corrosion profiles, and macrodefects [i.e., 49]. Two methods at room temperature that require point contact with the cold face of the furnace are known. The first is impact-echo method, used in construction concretes and pavements (Sect. 1.4). The second method is the frequency-modulated continuous-wave (FM-CW) radar technique [50], which can produce wall thickness data in real time.

Use of ultrasonic testing techniques has been attempted in harsh environments including high temperatures and radiation [i.e., 51]. Testing is complicated because wave-guides with special high temperature couplants and cooling systems are necessary to protect ultrasonic transducers from reaching their Curie point. Newer transducers, based on AlN films, capable of emitting and receiving ultrasonic energy at temperatures exceeding 900°C and pressures above 150 MPa have also been reported [52].

3.3.2 Moisture Profile

The measurement of moisture profiles while drying construction concretes has been reported using strain gauges on laminated specimens [53] and by using magnetic resonance imaging (MRI) [54]. This last procedure has been shown to determine moisture profiles nondestructively and with very high resolution, on the order of millimeter or less. Size of specimens is limited to small cylinders (~ 2.5 cm diameter) and in situ heating of specimens limits the technique to research applications. Electromagnetic modeling of the interaction of microwave signals with moist cement-based materials [55, 56] provides the necessary insight to evaluate water content distribution and movement in refractory castables in a nonintrusive manner and with potential high resolution (1 mm).

3.3.3 Castable Rheology

Refractory castable mixing is influenced by particle-size distribution (PSD) and the water addition method used. Castables require a minimum of mixing energy to reach maximum flow values, which is supplied by a two-step water-addition and can be designed with PSDs that result in high mixing efficiency combining low torque values, short mixing times, and controlled heating [57]. Rheological evaluation is accomplished with a novel rheometer used also for development of improved mixes [58]. Besides flowability, set time, and strength-gain measurements, exothermic profile measurements have also been used to assess rheology [59]. An exothermic profile on a neat cement paste (a mixture of water and hydraulic cement) gives information about the composition and reactivity, and it is suggested for use in quality control applications. Such applications include verifying setting time, troubleshooting problem castables, and screening reactive materials.

3.4 Current Research Topics

Refractory concretes are used today in a wide variety of industrial applications where pyro-processing or thermal containment is required [60]. Calcium aluminate cement (CAC)-based monolithics represent the most important chemical category, and the fundamental properties and behavior have been recently summarized (production, uses, CAC clinker, mineralogy, microstructure and hydration, and environmental issues) [61]. The general current working topics are depending on economic trends of the industry. The demand for refractories in the US is projected to increase 2.2% per year to $2.4 billion in 2007 [62]. The demand is expected to improve the steel market since steel manufacturing consumes about 60% of total refractories. Preformed shapes (shapes manufactured with refractory castables) and castables (for in situ monolithic manufacturing) are expected to be the fastest growth segment. Table 5 summarizes the expected refractory demand in the US.

A common trend seen today in this industry is plant integration through the merger and alliance of refractories companies. In general, the global refractory production has increased, due in part to process optimization (better usage of refractories) or to improved refractory materials with better performance. Monolithic refractories consumption, however, has markedly increased and represents today a large portion of the total. For instance, it is about 60% of the total production in Japan [63].

Current issues on refractory monolithics include no-cement castables, self-flow and free-flow mixes [64], and special compositions such as spinel-based [65], placement techniques [66], new applications (i.e., crown superstructures for glass melters [67]), characterization [68, 69], and basic castables [70, 71]. Specialized publications have compiled and reproduced most of the most important topics [i.e, 72–74], which include the following:

- Development of high performance castables (i.e., high-strength under corrosion conditions at high temperatures)
- Application, installation issues; drying and firing equipment; development of very large shapes

Table 5 Refractory product demand in the US (US$ million [62])

	1997	2002	2007
Total demand	2,518	2,180	2,430
Bricks and shapes	1,480	1,210	1,345
Monolithics	652	575	635
Other	386	395	450

- New binders for castables without affecting refractory properties; new dispersants and rheology studies
- Study of thermo-mechanical properties, chemical behavior, theoretical predictions, and computer simulation
- Development of basic castables: for instance, use of MgO, CaO, and dolomite for making high-temperature basic castables; coating of aggregates (i.e., silanes) for preserving clinkers
- Study and optimization of drying and firing: experimental and theoretical simulation of drying and its effects on mechanical strength and refractoriness.

References

1. The History of Concrete: http://matse1.mse.uiuc.edu/ tw/concrete/hist.html.
2. Ash Grove Cement Co.: http://www.ashgrove.com/careers/pdf/manufacturing.pdf.
3. J.A. Dobrowolski, *Concrete Construction Handbook*, 4th edn., NY, McGraw Hill, 1998.
4. World Cement and Concrete Additives to 2006, The Freedonia Group, Inc. Cleveland, OH, January 2003: http://www.freedoniagroup.com/.
5. D.J. Hannant, Fiber reinforcement in the cement and concrete industry, *Mater. Sci. Tech.*, 11(9) 853–861 (1995).
6. ASTM A 820: Specification for steel fibers for fiber reinforced concrete, The American Society for Testing and Materials, West Conshohocken, PA.
7. Report on Fiber Reinforced Concrete, ACI 544.1R-96, American Concrete Institute, Detroit, Michigan, http://www.concrete.org.
8. Report on Fiber Reinforced Plastic Reinforcement for Concrete Structures, ACI 440R-96, American Concrete Institute, Detroit, Michigan.
9. Saint-Gobain Vetrotex International: http://www.saint-gobainvetrotex.com/business_info/marketfig.html.
10. M.-T. Liang, P.-J. Su, Detection of the corrosion damage of rebar in concrete using impact-echo method, *Cem. Concr. Res.*, **31**, 1427–1436 (2001).
11. http://www.oit.doe.gov/inventions/factsheets/mosci.pdf: Low-Energy Alternative to Commercial Silica-Based Glass Fibers, MO-SCI Corporation Rolla, MO.
12. Blue Circle Rendaplus and Fibrocem GRC, Lafarge Cement, http://www.lafargecement.co.uk.
13. R.E. Green (ed.), Civil structures in nondestructive characterization of materials VII, *Proc. of the 8th International Symposium, Plenum Press*, NY, 1997, pp. 475–586.
14. J. Wollbold, Ultrasonic-impulse-echo-technique, advantages of an online-imaging technique for the inspection of concrete, www.ndt.net/abstract/ut97/civil497/neisec.htm.
15. M. Krause, Ch. Maierhofer, and H. Wiggenhauser, Thickness measurement of concrete elements using radar and ultrasonic impulse echo techniques, in *Proc. 6th International Conf. on Structural Faults and Repair*, Engineering Technics Press, London, 1995, pp. 17–24.
16. The Federal Institute for Materials Research and Testing (BAM), Unter den Eichen 87, 12205 Berlin, Germany, http://www.bam.de/.
17. M.J. Sansalone, and W.B. Street, *Impact-Echo: Nondestructive Evaluation of Concrete and Masonry*, Bullbrier Press, Ithaca, NY, 1997.
18. Report on Nondestructive Test Methods for Evaluation of Concrete in Structures, ACI 228.2R-98, American Concrete Institute, Detroit, Michigan.
19. M. Krause, M. Barmann, R. Frielinghaus, F. Kretzschmar, O. Kroggel, K.J. Langenberg, C. Maierhofer, W. Muller, J. Neisecke, M. Schickert, V. Schmitz, H. Wiggenhauser, and F. Wollbold, Comparison of pulse-echo methods for testing concrete, *NDT&E Int.*, 30(4), 195–204 (1997).
20. A.D. Davis, The nondestructive impulse response test in north america: 1985–2001, *NDT&E Int.* June 2003, 36[4], 185–193.
21. A. Sadri, Application of impact-echo technique in diagnoses and repair of stone masonry structures, *NDT&E Int.* (in press).

22. Y.S. Cho, Non-destructive testing of high strength concrete using spectral analysis of surface waves, *NDT&E Int.* (in press).
23. ASTM C 1383–98: Test Method for Measuring the P-Wave Speed and the Thickness of Concrete Plates using the Impact-Echo Method, The American Society for Testing and Materials, West Conshohocken, PA.
24. Y.S. Cho, and F.-B. Lin, Spectral analysis of wave response of multi-layer thin cement mortar slab structures with finite thickness, *NDT&E Int.*, **34**, 115–122 (2001).
25. J.S. Popovics, W. Song, J.D. Achenbach, J.H. Lee, and R.F. Andre, One-side stresswave velocity measurement in concrete, *J. Eng. Mech.*, 1346–1353 (1998).
26. M. Ohtsu, and T. Watanabe, Stack imaging of spectral amplitudes based on impact-echo for flaw detection," *NDT&E Int.*, **35**, 189–196 (2002).
27. K. Mori, A. Spagnoli, Y. Murakami, G. Kondo, and I. Torigoe, A new-non-contacting non-destructive testing method for defect detection in concrete, *NDT&E Int.*, **35**, 399–406 (2002).
28. ASTM C 947: Test Method for Flexural Properties of Thin-Section Glass-Fiber-Reinforced Concrete, The American Society for Testing and Materials, West Conshohocken, PA.
29. ASTM C 948: Test Method for Dry and Wet Bulk Density, Water Absorption, and Apparent Porosity of Thin Sections of Glass-Fiber Reinforced Concrete, The American Society for Testing and Materials, West Conshohocken, PA.
30. ASTM E 136: Test Method for Behavior of Materials in a Vertical Tube Furnace at 750°C, The American Society for Testing and Materials, West Conshohocken, PA.
31. ASTM E119: Test Methods for Fire Tests of Building Construction and Materials, The American Society for Testing and Materials, West Conshohocken, PA.
32. Center for Advanced Cement-Based Materials, Northwestern University: http://acbm.northwestern.edu/.
33. L.J. Struble (ed.), *Cements Research Progress 1995*, The American Ceramic Society, Westerville, OH, 1997.
34. A. Boyd, S. Mindess, and J. Skalny (eds.), *Materials Science of Concrete: Cement and Concrete – Trends and Challenges*, The American Ceramic Society, Westerville, OH, 2002.
35. Vision 2030: A Vision for the U.S. Concrete Industry, December 2002: http://www.concretesdc.org/concrete_roadmap.pdf.
36. Concrete Projects for Oil Industry, Heidrun & Troll, Norway, http://www.international.ncc.se/about_international/Heidrun.pdf.
37. D.J. Naus, Activities under the Concrete and Containment Technology Program at ORNL, Technical Report ORNL/TM-2002/213; http://www.osti.gov/bridge.
38. ISO 836:2001: Terminology for refractories, International Organization for Standardization, Geneva, Switzerland.
39. ISO 1927:1984: Prepared unshaped refractory materials (dense and insulating), International Organization for Standardization, Geneva, Switzerland.
40. Technology of Monolithic Refractories, Chap. 1 Revised edition, Plibrico Japan Co., Ltd., Tokyo Insho Kan Printing Co., Ltd., 1999.
41. G. Routschka (ed.), *Refractory Materials*, Chap. 5, Vulkan-Verlag, Essen, Germany, 1997.
42. S. Banerjee, *Monolithic Refractories*, Chap. 2, The American Ceramic Society/World Scientific, 1998.
43 M. Velez, A. Erkal, and R.E. Moore, Computer simulation of the dewatering of refractory concrete walls, *J. Tech. Assoc. Refractories* (Taikabutsu Overseas), **20**(1), 5–9 (2000).
44. G. Routschka, J. Potschke, and M. Ollig, Gas Permeability of Refractories at Elevated Temperatures, *Proc. UNITECR'97*, 1997, pp. 1550–1565.
45. M.D.M. Innocentini, M.G. Silva, B.A. Menegazzo, and V.C. Pandolfelli, Permeability of refractory castables at high temperatures, *J. Am. Ceram. Soc.*, **84**(3), 645–647 (2001).
46. Z.-X. Gong, and A.S. Mujumdar, Development of drying schedules for one-side-heating drying of refractory concrete slabs based on a finite element model, *J. Am. Ceram. Soc.*, **79**(6) 1649–1658 (1996).
47. W.L. Headrick, Quality time – automated equipment helps achieve rapid determination of gas permeability, *Ceram. Ind.*, **148**(2) 74–76 (1998).
48. R.E. Moore, J.D. Smith, and T.P. Sander, Dewatering monolithic refractory castables: experimental and practical experience, *Proc. UNITECR'97*, 1997, pp. 573–583.

49. D.A. Bell, A.D. Deighton, and F.T. Palin, Non-destructive testing of refractories, in advances in refractories for the metallurgical industries II, *Proc. Int. Symposium of the Canadian Institute of Mining, Metallurgy and Petroleum*, Montreal, Quebec, August 24–29, 1996, pp. 191–207.

50. R. Zoughi, L.K. Wu, and R.K. Moore, SOURCESCAT: a very-fine-resolution radar scatterometer, *Microwave J.*, **28**(11) 183–196 (1985).

51. B. Audoin, and C. Bescond, Measurement by laser-generated ultrasound of four stiffness coefficients of an anisotropic material at elevated temperatures, *J. Nondestructive Eval.* **16**(2) 91–100 (1997).

52. R.E. Dutton, and D.A. Stubbs, An ultrasonic sensor for high temperature materials, in *Sensors and Modeling in Materials Processing: Techniques and Applications*, S. Viswanathan, R.G. Reddy, and J.C. Malas (eds.), The Minerals, Metals and Materials Society, 1997, pp. 295–303.

53. G.L. England, and N. Khoylou, Modeling of moisture behavior I normal and high performance concretes at elevated temperatures, in *4th International Symposium on the Utilization of High Strength/High Performance Concrete*, May 29–31, 1995, pp. 53–68.

54. S.D. Beyea, B.J. Balcom, T.W. Bremmer, P.J. Prado, D.P. Green, R.L. Armstrong, and P.E. Grattan-Bellew, Magnetic resonance imaging and moisture content profiles of drying concretes, *Cem. Concr. Res.*, **28**(3) 453–463 (1998).

55. S. Peer, Nondestructive evaluation of moisture and chloride ingress in cement-based materials using near-field microwave techniques, M.S. Thesis, Electrical and Computer Engineering, University of Missouri-Rolla, MO, December 2002.

56. S. Peer, K.E. Kurtis, and R. Zoughi, An electromagnetic model for evaluating temporal water distribution and movement in cyclically soaked mortar, *IEEE Transactions* on *Instrumentation and Measurement*, April 2004, V.53[2], 406–415.

57. R.G. Pileggi, A.R. Studart, V.C. Pandolfelli, and J. Gallo, How mixing affects the rheology of refractory Castables, *Am. Ceram. Soc. Bull.*, **80**(6) 27–31 (2001); **80**(7) 38–42 (2001).

58. A.R. Studart, R.G. Pileggi, V.C. Pandolfelli, and J. Gallo, High-alumina multifunctional refractory castables, *Am. Ceram. Soc. Bull.*, **80**(11) 34–40 (2001).

59. C. Alt, L. Wong, and C. Parr, Measuring castable rheology by exothermic profile, *Refract. Appl. News* **8**(2) 15–18 (2003).

60. Refractories, The Freedonia Group, Inc., Cleveland, OH, August 2003, http://www.freedonia-group.com/.

61. Refractory Concrete: State-of-the-Art Report, ACI 547.1R-87, American Concrete Institute, Detroit, Michigan, http://www.concrete.org.

62. R.J. Magabhai, and F. Glasses (eds.), Calcium Aluminate Cements, *Proc. Int. Symposium on Calcium Aluminate Cements*, Edinburgh, Scotland, Maney Publications, UK, July 2001.

63. K. Murakami, T. Yamato, Y. Ushijima, and K. Asano, The trend of monolithic refractory technology in Japan, *Refract. Appl. News*, **8**(5) 12–16 (2003).

64. P.C. Evangelista, C. Parr, and C. Revais, Control of formulation and optimization of self-flow castables based on pure calcium aluminates, *Refract. Appl. News*, **7**(2) 14–18 (2002).

65. T.A. Bier, C. Parr, C. Revais, and M. Vialle, Spinel forming castables: physical and chemical mechanisms during drying, *Refract. Appl. News* **5**(4) 3–4 (2000).

66. T. Richter, and D. McIntyre, Novel form free installation method for refractory castables, *Refract. Appl. News* **6**(4) 3–5 (2001).

67. M. Velez, Magneco/Metrel, an evolution in refractory monolithics, *Refract. Appl. News* **6**(1) 14 (2001).

68. W.L. Headrick, and R.E. Moore, Sample preparation, thermal expansion, and Hasselman's thermal shock parameters of self-flow refractory castables, *Refract. Appl. News* **7**(1) 9–15 (2001).

69. W.E. Lee, S. Zhang, and H. Sarpoolaky, Different types of in situ refractories, *Refract. Appl. News* **6**(2) 3–4 (2001).

70. S.L.C. da Silva, Improvement of the hydration resistance of magnesia and doloma using organosilicon compounds, Ph.D. Thesis, UMR, Ceramic Engineering Dept., 2000.

71. U.S. Patent 5,183,648, Process for preparing magnesia having reduced hydration tendency and magnesia based castable, Assigned to Shell Research Ltd., 1993.

72. Refractories Applications & News, published by the University of Missouri-Rolla: http://www.ranews.info.

73. *The* Refractories Engineer, published by The Institute of Refractories Engineers (UK): http://ireng.org/.
74. Taikabutsu, published by the Technical Association of Refractories of Japan: http://www.tarj.org/.

Relevant ASTM standards

1. ASTM C 39: Test Method for Compressive Strength of Cylindrical Concrete Specimens, February 2001.
2. ASTM C 78: Test Method for Flexural Strength of Concrete, January 2002.
3. ASTM C 157: Test Method for Length Change of Hardened Hydraulic-Cement Mortar and Concrete, July, 1999.
4. ASTM C 231: Test Method for Air Content of Freshly Mixed Concrete by the Volumetric Method, March 2001.
5. ASTM C 417: Standard Test Method for Thermal Conductivity of Unfired Monolithic Refractories, December 1993, Revised 1998.
6. ASTM C860: Standard Practice for Determining the Consistency of Refractory Castable Using the Ball-in-Hand Test, April 2000.

Chapter 9
Lead Compounds

Julie M. Schoenung

Abstract Lead compounds include over forty naturally occurring minerals from which five lead oxides can be derived. The lead oxides, as well as some lead silicates, are used as raw materials in lead-containing glasses and crystalline electronic ceramics. The presence of lead in glass increases the refractive index, decreases the viscosity, increases the electrical resistivity, and increases the X-ray absorption capability of the glass. The lead in electronic ceramics increases the Curie temperature and modifies various electrical and optical properties. The refinement of metallic lead from minerals and recycled goods such as lead acid batteries and cathode ray tubes is a multistep process, supplemented by oxidation steps to produce lead oxides. Lead compounds are known to be toxic and are therefore highly regulated.

1 Introduction

Lead and lead compounds have been used in a multitude of products for centuries. Lead (metal) is occasionally used as a "pure" material, but this is relatively rare when compared with the extent of its use in alloys and in ceramic compounds and glasses.

Lead is the 82nd element in the periodic table. It is present in the IVA column below carbon, silicon, germanium, and tin, and in the sixth row between thallium and bismuth. It is metallic in its pure state and crystallizes into the face-centered-cubic crystal structure. Lead has a low bond energy, as is evidenced by its low melting point (327°C). Lead and its alloys exhibit low elastic moduli, yield strength, and tensile strength when compared with other metals, glasses, and technical ceramics (see Table 1). The fracture toughness is also low when compared with other metals. The lead atom is large (atomic radius = 0.175 nm) and exhibits two possible oxidation states: +2 and +4. Lead is one of the commonly used heaviest metals with an atomic weight of 207.2 amu and a density of bulk material 11.35 g cm^{-3} at 20°C. These fundamental, chemical, and physical attributes define the foundation for the reason why lead is used in most of its applications. The most common and important applications of lead and lead compounds in ceramics and glasses are described in Sect. 2.

J.F. Shackelford and R.H. Doremus (eds.), *Ceramic and Glass Materials:*
Structure, Properties and Processing.
© Springer 2008

Table 1 Selected mechanical properties of various materials [1]

Material	Elastic modulus (GPa)	Yield strength (MPa)	Tensile strength (MPa)	Fracture toughness (MPa m$^{1/2}$)
Lead alloys	12.5–15.0	8–14	12–20	5–15
Aluminum alloys	68–82	30–500	58–550	22–35
Copper alloys	112–148	30–500	100–550	30–90
Iron alloys	165–217	170–1,155	345–2,240	12–280
Glasses	61–110	264–2,129[a]	22–177	0.5–1.7
Technical ceramics	140–720	524–6,833[a]	160–800	0.8–6.0
Leather	0.1–0.5	5–10	20–26	3–5
Polyethylene	0.6–0.9	18–29	21–45	1.4–1.7
Polypropylene	0.9–1.6	21–37	28–41	3.0–4.5
Polyvinylchloride	2.1–4.1	35–52	41–65	1.5–5.1

[a] Yield strength for glasses and ceramics is measured in compression; all other materials are measured in tension

Lead is found in a wide variety of naturally occurring minerals (see Table 2). These minerals range from rather simple substances, such as pure lead, PbTe, PbSe, and PbS, to complex hydroxides, such as $Pb_2Cu(AsO_4)(SO_4)OH$ and $Pb_{26}Cu_{24}Ag_{10}Cl_{62}(OH)_{48}$ $3H_2O$. As shown in Table 3, these minerals represent a wide range of crystal systems, of which the most common are monoclinic, orthorhombic, and tetragonal. Low hardness values (typically between 2 and 3 Mohs with extreme values of 1.5 for pure lead and 5.5 for plattnerite) and high theoretical densities (typically greater than 5 and as high as 11.3 g cm^{-3}) are characteristic of these lead-containing minerals.

As described in Sect. 3, these minerals can be refined to produce metallic lead, or they can be processed to produce lead oxides. Because of lead's two oxidation states, four lead oxide compositions are possible: PbO, PbO_2, Pb_2O_3, and Pb_3O_4. The PbO composition can form into two different crystal structures: orthorhombic (called massicot) and tetragonal (called litharge). Thus, five possible lead-containing oxides are available for glass and ceramic fabrication. The JCPDS cards that describe the crystallographic characteristics for these oxides are as follows: 05-0561 for litharge (PbO), 38-1477 for massicot (PbO), 41-1492 for platnerite (PbO_2), and 41-1494 for minium (Pb_3O_4). Litharge is the most commonly used oxide for glass and ceramic fabrication. Alternatively, lead silicates can also be used. These include ($2PbO–SiO_2$), ($PbO–SiO_2$), and ($4PbO–SiO_2$). Selected physical, thermal, and mechanical properties of the lead oxides are listed in Table 4. It can be seen that for all of these oxides, the lead content is very high (85–93 wt%), the density is high (8.9–10.1 gcm^{-3}), and the hardness values are low (2–2.5 Mohs). The melting point values show more variability, ranging from 290 to 888°C. Thermodynamic data for the lead oxides, lead silicates, and selected lead-containing minerals are presented in Table 5.

Many lead-containing products, including leaded glass, can be recycled and provide another source of material to supplement the naturally occurring minerals. The processing required to produce metallic lead and lead oxides are outlined in Sect. 3. Descriptions of the most important sources of lead and statistics on lead production and consumption are also presented.

The use of lead and lead compounds, although ubiquitous at present, is expected to decrease in the future because of health concerns. It is commonly known that lead is toxic to humans, especially children. As a consequence, legislative bodies have

Table 2 Various lead-containing minerals [2]

Mineral	Chemical name	Chemical formula
Altaite	Lead telluride	$PbTe$
Anglesite	Lead sulfate	$PbSO_4$
Arsentsumebite	Lead copper arsenate sulfate hydroxide	$Pb_2Cu(AsO_4)(SO_4)OH$
Baumhauerite	Lead arsenic sulfide	$Pb_3As_4S_9$
Bayldonite	Hydrated copper lead arsenate hydroxide	$Cu_3Pb(AsO_4)_2 \cdot H_2O$
Beudantite	Lead iron arsenate sulfate hydroxide	$PbFe_3AsO_4SO_4(OH)_6$
Bideauxite	Lead silver chloride fluoride hydroxide	$Pb_2AgCl_3(F,OH)_2$
Bindheimite	Lead antimony oxide hydroxide	$Pb_2Sb_2{}_6(O,OH)$
Boleite	Hydrated lead copper silver chloride hydroxide	$Pb_{26}Cu_{24}Ag_{10}Cl_{62}(OH)_{48} \cdot 3H_2O$
Boulangerite	Lead antimony sulfide	$Pb_5Sb_4S_{11}$
Caledonite	Copper lead carbonate sulfate hydroxide	$Cu_2Pb_5CO_3(SO_4)_3(OH)_6$
Cerussite	Lead carbonate	$PbCO_3$
Clausthalite	Lead selenide	$PbSe$
Crocoite	Lead chromate	$PbCrO_4$
Cumengite	Lead copper chloride hydroxide	$Pb_{21}Cu_{20}Cl_{42}(OH)_{40}$
Diaboleite	Copper lead chloride hydroxide	$CuPb_2Cl_2(OH)_4$
Dundasite	Hydrated lead aluminum carbonate hydroxide	$Pb_2Al_4(CO_3)_4(OH)_8 \cdot 3H_2O$
Fiedlerite	Lead chloride fluoride hydroxide	$Pb_3Cl_4F(OH)_2$
Galena	Lead sulfide	PbS
Gratonite	Lead arsenic sulfide	$Pb_9As_4S_{15}$
Hedyphane	Lead calcium arsenate chloride	$Pb_3Ca_2(AsO_4)_3Cl$
Jordanite	Lead arsenic antimony sulfide	$Pb_{14}(As,Sb)_6S_{23}$
Laurionite	Lead chloride hydroxide	$PbClOH$
Leadhillite	Lead sulfate carbonate hydroxide	$Pb_4SO_4(CO_3)_2(OH)_2$
Massicot	Lead oxide	PbO
Meneghinite	Lead antimony sulfide	$Pb_{13}Sb_7S_{23}$
Mimetite	Lead chloroarsenate	$Pb_5(AsO_4)_3Cl$
Minium	Lead oxide	Pb_3O_4
Native lead	Elemental lead	Pb
Nealite	Lead iron arsenate chloride	$Pb_4Fe(AsO_4)_2Cl_4$
Phosgenite	Lead carbonate chloride	$Pb_2CO_3Cl_2$
Plattnerite	Lead oxide	PbO_2
Pseudoboleite	Hydrated lead copper chloride hydroxide	$Pb_5Cu_4Cl_{10}(OH)_8 \cdot 2H_2O$
Pyromorphite	Lead chlorophosphate	$Pb_5(PO_4)_3Cl$
Semseyite	Lead antimony sulfide	$Pb_9Sb_8S_{21}$
Susannite	Lead sulfate carbonate hydroxide	$Pb_4SO_4(CO_3)_2(OH)_2$
Vanadinite	Lead chlorovanadinate	$Pb_5(VO_4)_3Cl$
Wulfenite	Lead molybdenate	$PbMoO_4$

targeted the use of lead in numerous products, mandating labeling, recycling, and/or complete termination of use. The known health risks and existing legislative initiatives dealing with lead and lead compounds are summarized in Sect. 4.

Table 3 Crystal structure, hardness, and density for various lead-containing minerals [2]

Mineral	Crystal system	Hardness (Mohs)	Density (g cm^{-3})
Altaite	Isometric	2.5–3	8.2–8.3
Anglesite	Orthorhombic	2.5–3.0	6.3+
Arsentsumebite	Monoclinic	3	6.4
Baumhauerite	Triclinic	3	5.3
Bayldonite	Monoclinic	4.5	5.5
Beudantite	Rhombohedrons, pseudocubic	4	4.3–4.5
Bideauxite	Isometric	3	6.3
Bindheimite	Isometric	4–4.5	7.3–7.5
Boleite	Tetragonal	3–3.5	5+
Boulangerite	Monoclinic	2.5	5.8–6.2
Caledonite	Orthorhombic	2.5–3	5.6–5.8
Cerussite	Orthorhombic	3.0–3.5	6.5+
Clausthalite	Isometric	2.5	8.1–8.3
Crocoite	Monoclinic	2.5–3	6.0+
Cumengite	Tetragonal	2.5	4.6
Diaboleite	Tetragonal	2.5	5.4–5.5
Dundasite	Orthorhombic	2	3.5
Fiedlerite	Monoclinic	3.5	5.88
Galena	Cubic and octahedron	2.5+	7.5+
Gratonite	Trigonal	2.5	6.2
Hedyphane	Hexagonal	4.5	5.8–5.9
Jordanite	Monoclinic	3	5.5–6.4
Laurionite	Orthorhombic	3–3.5	6.1–6.2+
Leadhillite	Monoclinic	2.5–3	6.3–6.6
Massicot	Orthorhombic	2	9.6–9.7
Meneghinite	Orthorhombic	2.5	6.3–6.4
Mimetite	Hexagonal	3.5–4	7.1+
Minium	Tetragonal	2.5–3	8.9–9.2
Native lead	Isometric	1.5	11.3+
Nealite	Trigonal	4	5.88
Phosgenite	Tetragonal	2.0–3.0	6.0+
Plattnerite	Tetragonal	5–5.5	6.4+
Pseudoboleite	Tetragonal	2.5	4.9–5.0
Pyromorphite	Hexagonal	3.5–4	7.0+
Semseyite	Monoclinic	2.5	5.8–6.1
Susannite	Trigonal	2.5–3	6.5
Vanadinite	Hexagonal	3	6.6+
Wulfenite	Tetragonal	3	6.8

Table 4 Selected physical, thermal, and mechanical properties of various lead oxides

Oxide	Formula weight (g mol^{-1})	Lead content (wt%)	Crystal system	Density (g cm^{-3})	Melting point (°C)	Hardness (Mohs)
PbO (massicot)	223.2	92.8	Orthorhombic	9.64	489	2
PbO (litharge)	223.2	92.8	Tetragonal	9.35	888	2
PbO$_2$	239.2	86.6	Tetragonal	9.64	290	5.5
Pb$_2$O$_3$	462.4	89.6	Monoclinic	10.05	530[a]	
Pb$_3$O$_4$	685.6	90.1	Tetragonal	8.92	830	2.5

[a] Decomposition temperature

Table 5 Thermodynamic properties of various lead-containing minerals, lead silicates, and lead oxides [3]

Chemical formula	$\Delta_f H^o$ (kJ mol^{-1})	$\Delta_f G^o$ (kJ mol^{-1})	S^o (J (mol K)$^{-1}$)	C_p (J (mol K)$^{-1}$)
PbTe	−70.7	−69.5	110.0	50.5
PbS	−100.4	−98.7	91.2	49.5
PbSe	−102.9	−101.7	102.5	50.2
PbCO$_3$	−699.1	−625.5	131.0	87.4
PbSO$_4$	−920.0	−813.0	148.5	103.2
PbCrO$_4$	−930.9			
PbMoO$_4$	−1,051.9	−951.4	166.1	119.7
PbSiO$_3$	−1,145.7	−1,062.1	109.6	90.0
Pb$_2$SiO$_4$	−1,363.1	−1,252.6	186.6	137.2
PbO (Massicot)	−217.3	−187.9	68.7	45.8
PbO (Litharge)	−219.0	−188.9	66.5	45.8
PbO$_2$	−277.4	−217.3	68.6	64.6
Pb$_3$O$_4$	−718.4	−601.2	211.3	146.9

$\Delta_f H^o$ standard molar enthalpy (heat) of formation at 298.15 K in kJ mol^{-1}; $\Delta_f G^o$ standard molar Gibbs free energy of formation at 298.15 K in kJ mol^{-1}; S^o standard molar entropy at 298.15 K in J (mol K)$^{-1}$; C_p molar heat capacity at constant pressure at 298.15 K in J (mol K)$^{-1}$

2 Applications

Lead is one of the most widely used substances in the world, with applications as a pure metal, as an alloying element in other metals, as an additive in organic substances, and as an additive or primary material component in ceramics and glasses. Lead, in metallic form is used in numerous applications, including lead-acid batteries, lead sheet and pipe, sheathing for electrical cable, radiation shielding, and lead shot and weights. As an alloying element, lead is used extensively in lead–tin solders for electronic packaging and other applications. Lead is also an alloying element in bronzes, steels, and aluminum alloys. As an additive in organic substances, lead is used in pigments, paints, polymers, and gasoline. The focus of the remainder of this section, however, is the use of lead in making ceramics and glasses.

The applications for lead and lead compounds, mostly oxides, as used in ceramic and glass applications can be categorized as follows:

Glasses

1. Leaded glass ("crystal") for household products
2. Glazes and enamels for ceramic whitewares
3. High-index optical and ophthalmic glass
4. Radiation shielding glass
5. High electrical resistance glass for lamps and display technologies
6. Glass solders and sealants for glass-to-glass joining and hermetic glass-to-metal sealing

Electronic ceramics

1. Capacitor dielectrics
2. Piezoelectrics
3. Electrooptic devices

The primary reasons for adding lead to glass are to increase the refractive index of the glass, to decrease the viscosity of the glass, to increase the electrical resistivity of the glass, and to increase the X-ray absorption capability of the glass used for radiation shielding. The primary reason for using lead-based electronic ceramics is to modify the dielectric and piezoelectric properties, such as Curie point and piezoelectric coupling factor.

There are numerous glass products that contain lead. Because lead has two oxidation states (+2 and +4), the lead in glass can act as either a network former by replacing the silicon atom, or a network modifier by causing the formation of nonbridging oxygen atoms [4, 5], as shown in Fig. 1. The presence of lead breaks up the Si–O network and significantly reduces the viscosity of the glass (see Fig. 2). The working point of a high-lead glass, for instance, is reduced to approximately 850°C, compared to ~1,100°C for soda lime glass and >1,600°C for fused silica.

Leaded glass, which is used in houseware applications such as decorative glassware and vases, is commonly (and erroneously) referred to as "crystal" because it exhibits a higher index of refraction than other glasses. Representative values of the index of refraction for various glasses are listed in Table 6. This property results in the glass appearing shinier, brighter, and more colorful than a typical glassware (soda lime silica) glass. Leaded glass for these applications typically use PbO as a raw material, with content ranging from 18 to 38 wt% PbO [10]; a representative value is 24.4 wt% PbO [11].

Glazes for ceramic bodies and porcelain enamels for metallic substrates are coatings that are applied to these surfaces with a variety of purposes: chemical inertness, zero permeability to liquids and gases, cleanability, smoothness and resistance to abrasion and scratching, mechanical strength, and decorative and aesthetic considerations [12].

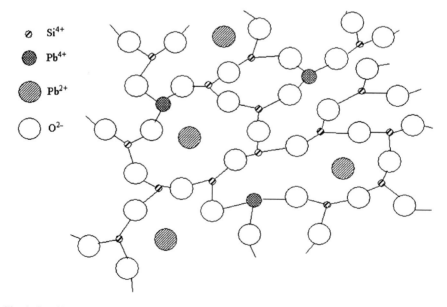

Fig. 1 Lead in glass, acting as either a network former or network modifier [6]

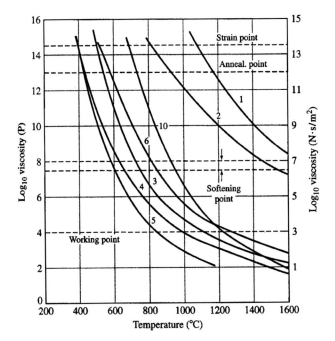

Fig. 2 Viscosity versus temperature characteristics for various glass compositions [7]. 1, Fused silica; 2, 96% silica; 3, soda lime (plate glass); 4, lead silicate (electrical); 5, high-lead; 6, borosilicate (low expansion); 10. aluminosilicate

Table 6 Refractive indices of various glasses [8,9]

Glass composition	Average refractive index
Silica glass, SiO_2	1.458
Vycor glass (96% SiO_2)	1.458
Soda lime silica glass	1.51–1.52
Borosilicate glass (Pyrex ™)	1.47
Dense flint optical glasses	1.6–1.7
Lead silicate glasses	2.126

These coatings are applied to numerous products including china, vases, sinks, toilets, and washing machines. Lead, in the form of litharge (PbO), is often added to glazes, but not usually to enamels, because it reduces the viscosity of the glass, which in turn provides a smoother, more corrosion-resistant surface. The higher index of refraction that results is desirable for these applications. Lead-containing glazes typically have a composition between 16 and 35wt% PbO [13–15].

Optical glass includes a wide variety of applications. Of these, lead oxide is most often incorporated into optical flints, although it might also be added to optical crown glass, ophthalmic glass (crown or flint), and optical filter glass [16]. For example, products in which the presence of lead is valued include Cerenkov counters, magnetooptical switches and shutters, and the cores of fiberoptic faceplates [17]. One of the reasons why lead is added to optical glass is it creates a high index of refraction, which can facilitate total internal reflection. The lead content of optic glasses varies considerably:

6–65 wt% PbO in optical flint glass, 4 wt% PbO in optical crown glass, and 6–51 wt% in ophthalmic glass [4,16].

Radiation shielding glass is used in television and computer monitors that contain cathode ray tubes (CRTs) because CRTs generate X-rays [1]. Exposure of the viewer to these X-rays is undesirable and limited by US Federal Standard Public Law 90–602 (Radiation Control for Health and Safety Act, 1968). X-ray absorption by a given material is dependent upon the wavelength of the radiation and the density, thickness, and atomic number of the material. Because a lead-free glass might exhibit a linear absorption coefficient as low as 8.0 cm^{-1} [18], lead is often added to CRT glass to provide the required X-ray shielding. The primary glass components of the CRT include the panel (or faceplate), the funnel, and the neck. Representative lead compositions and linear absorption coefficients, for the corresponding glass components, are shown in Table 7.

As a result of the large ionic size of the Pb^{2+} ion (0.132 nm), the electrical resistivity of leaded glass is orders of magnitude higher than that of lead-free, soda lime glass (direct-current (DC) resistivity at 250°C: 10$^{8.5}$ and 10$^{6.5}$ ohm-cm, respectively [19]) [20,21]. This characteristic of leaded glass is a primary reason why it is used for the stem and exhaust tube in many light fixtures: incandescent, fluorescent, and high-pressure mercury fixtures, as well as for hermetic seals in electronic devices. A typical composition for leaded glass in lamps is 20–22 wt% PbO [19,22].

As discussed earlier, the presence of lead in glass results in a significant change in its viscosity characteristics (see Fig. 2). Although this is true for all of the lead-containing glasses discussed earlier, it is of particular significance for the applications of glass solders (for joining glass to glass) and sealants (for joining glass to metal), which are almost always made from high-lead glasses. For instance, leaded glass is used to join the panel of a CRT to the funnel, to seal electronic packages, to bond the recording head, and to seal the panel on a flat panel display. Compositions for these high-lead glasses range from 56–77 wt% PbO, with the higher values being more common [4,18,23].

Typical PbO content for the lead-containing glass products described earlier are summarized in Table 8.

Crystalline, lead-containing ceramics generally fall within the category of materials called PZTs/PLZTs, which are lead-(lanthanum) zirconate titanates. These materials

Table 7 PbO content and linear absorption coefficient requirements in CRT components for color monitors [18]

Component	PbO content (wt%)	Linear absorption coefficient (cm^{-1})
Panel	2.2	28.0
Funnel	23.0	62.0
Neck	28.0	90.0

Table 8 Summary of lead oxide content in various glass products

Product	PbO content (wt%)
Leaded glass ("crystal") for household products	18–38
Glazes and enamels for ceramic whitewares	16–35
High-index optical and ophthalmic glass	4–65
Radiation shielding glass	2–28
High electrical resistance glass	20–22
Glass solders and sealants	56–77

are ferroelectrics, with the perovskite ($CaTiO_3$) crystal structure and unusual dielectric properties [24,25]. They are used in capacitor dielectrics, piezoelectrics, and electrooptic devices [26–28]. For capacitor applications, important properties include dielectric constant, capacitance deviation, and maximum dissipation factor. Lead-based compositions for capacitor dielectrics include lead titanate, lead magnesium niobate, lead zinc niobate, and lead iron niobate–lead iron tungstate. For piezoelectric applications such as sensors and actuators, important properties include electromechanical coupling factors, piezoelectric constants, permittivity, loss tangent, elastic constants, density, mechanical quality factor, and Curie temperature. Lead-based compositions for piezoelectrics generally fall into the PZT category [$Pb(Zr,Ti)O_3$], although proprietary compositions generally include dopants such as Mg, Nb, Co, Ni, Mo, W, Mn, Sb, and Sn. For electrooptic applications, important properties include optical transmittance and haze; linear electrooptic effect coefficient (r_c), second-order (or quadratic) electrooptic effect coefficient (R), and half-wave voltage; dielectric constant, ferroelectric hysteresis loop characteristics, and piezoelectric coupling constants; and microstructure, grain size, and porosity. The general composition for electrooptic devices is PLZT: $Pb_{1-x}La_x(Zr_yTi_{1-y})_{1-x/4}O_3$. The compositions for all three of these product applications represent a lead content on the order of 55–70 wt%.

3   Processing

Historically, in the United States, the consumption of lead in glasses and ceramics has been approximately 30,000–50,000 metric tons per year, which represents 2–3% of the total U.S. annual lead consumption [29]. If storage battery usage is not included in the annual total, as this product category represents over 86% of U.S. lead consumption annually, then glasses and ceramics represent 13–22% of the remaining demand for lead in the United States.

Litharge and the other lead oxides that are used in the production of glasses and ceramics are obtained primarily through the oxidation of refined (purified) metallic lead. Because metallic lead does not occur naturally in large quantities, it must be extracted from either primary sources (mineral ores) or secondary sources (recycled materials such as lead-acid batteries and cathode ray tubes). The processing required to refine metallic lead can be broken down into three major steps, as seen in Fig. 3:

1. Mining and concentrating
2. Extraction or smelting
3. Refining

The refining step is then followed by an oxidation step in order to produce lead oxide. Because these processes are discussed in detail in several other sources [1,30–32], the description provided below is intentionally brief.

For primary sources of lead, namely mineral ores, the process of mining and concentrating, indeed, begins at a mine. For secondary sources, this stage of the process is replaced by separation and sorting steps to remove the components in the batteries and CRTs that do not contain lead. The remaining process steps are fundamentally the same.

Considering the primary sources of lead, although there are over forty different minerals that contain lead (see Table 2), the three most common minerals from which

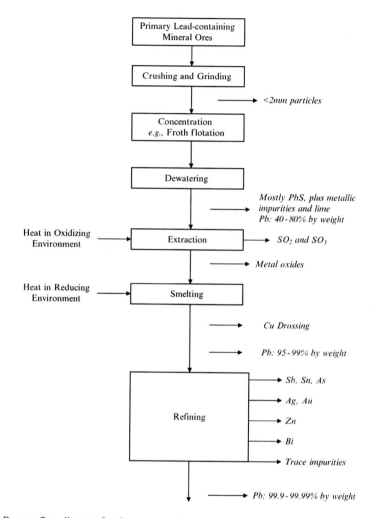

Fig. 3 Process flow diagram for the concentration, extraction, and refining of lead

pure lead is derived are galena (PbS), anglesite (PbSO$_4$), and cerussite (PbCO$_3$) with lead concentrations (by weight) of 87%, 68%, and 77.5%, respectively. Galena is easily recognized in the field because of its characteristic cubic shape, metallic luster, and high density. Anglesite and cerussite result from the natural weathering of galena. These three minerals exhibit the rock salt (NaCl), barite (BaSO$_4$), and aragonite (CaCO$_3$) crystal structures, respectively. The JCPDS cards that describe the crystallographic characteristics for these minerals are as follows: 05–0592 for galena (PbS), 36–1461 for anglesite (PbSO$_4$), and 47–1734 for cerussite (PbCO$_3$).

After being mined, these lead-containing minerals proceed through a concentration process that increases the lead concentration and removes waste (non-galena) rock, which is called gangue. The concentration process generally begins with crushing and grinding steps that ultimately result in particles <2 mm in size, followed by the actual concentration step, which is sometimes referred to as "beneficiation." The most common method used is froth flotation in which particles are separated on the basis of specific gravity. Additives, such as conditioners, are used to facilitate the separation process. Dewatering is then required, and the product remaining is primarily PbS. The concentration of lead in this product is generally between 40 and 80 wt%. Other substances contained in the material at this point include iron (Fe), zinc (Zn), copper (Cu), antimony (Sb), arsenic (As), silver (Ag), gold (Au), bismuth (Bi), and lime (CaO).

The extraction, or smelting, stage requires multiple steps. First, the lead concentrate is heated to remove the sulfur (as SO_2 and SO_3) and to agglomerate the fine particles. This step leads to sintering and oxidation of the material, thus forming oxides of lead, zinc, iron, and silicon. Other substances such as lime, metallic lead, and residual sulfur might also be present. The next step is the actual smelting, in which the material is heated in a reducing environment so that the oxides are converted into molten metals that can be refined and separated. Various contaminants are also removed through combustion. The lead at this point is called bullion and has a concentration between 95 and 99% lead by weight. Prior to actual refining, there is one more step, the copper drossing step, which is yet another heat treatment by which copper is removed.

The lead refining stage can be divided into five steps, each of which is designed to remove (and collect) selected impurities. The impurities removed during each step are as follows:

1. Antimony (Sb), tin (Sn), and arsenic (As)
2. Silver (Ag) and gold (Au)
3. Zinc (Zn)
4. Bismuth (Bi)
5. Trace impurities

After these steps, the lead concentration is now between 99.9 and 99.99 wt%.

It should be noted that, although existing methods are well established, new methods for smelting and refining continue to be developed. For instance, "direct" smelters are available, which eliminate the need for sintering, and electrolytic refining can be used as a one-step method for simultaneously removing all impurities (except tin). Each method presents its own combination of product quality, process cost, and environmental management requirements.

Lead-containing glasses and ceramics do not use metallic lead as a raw material. Instead, lead oxides and lead silicates are used. Various processing techniques are used to produce the lead oxides from refined metallic lead. For instance, litharge (PbO) is the reaction product of lead and oxygen, and can thus be produced by heating lead in air or by blowing air into molten lead. Minium (lead tetroxide), which is more oxygen-rich than litharge, can be created by further oxidizing litharge in a controlled atmosphere at about 450°C. Lead silicates are made by mixing and heating litharge and sand (SiO_2).

Primary and secondary sources of lead exist in the United States and throughout the world [29]. U.S. mine production of lead in concentrate is approximately 450,000–500,000 metric tons per year, which represents approximately 15% of the world production. Other countries with significant mine production of lead include Australia, Canada, China, Mexico, and Peru. Refining of secondary lead is dominated by the U.S. production, although other major sources include Canada, China, France, Germany, Italy, Peru, Spain, and the United Kingdom. In the United States, approximately 79% of the current lead refinery production is derived from secondary sources. Worldwide, secondary sources account for approximately 45% of the total 6.4 million metric tons of lead refined each year. Current prices for metallic lead and for litharge (PbO) are approximately \$0.95/kg and \$0.55/kg, respectively [33,34].

4 Health Effects and Legislation

Lead is ranked second on the CERCLA (Comprehensive Environmental Response, Compensation and Liability Act, 1980) Priority List of Hazardous Substances [35], indicating the national concern over lead with respect to its frequency of use, toxicity, and potential for human exposure. It is also one of the most highly regulated substances in the world. As a consequence, lead has been and will continue to be studied and monitored with respect to its human and ecological toxicity. At issue, primarily, is lead's human toxicity.

Lead can affect almost every organ and system in [the human] body. The most sensitive is the central nervous system, particularly in children. Lead also damages kidneys and the reproductive system [36].

The reader is referred to *Toxicological Profile for Lead* [37] for a comprehensive report on lead, including public health statement; health effects summary; chemical and physical information; production, import, use, and disposal information; description of the potential for human exposure; analytical methods used in monitoring health effects; regulations and advisories; a valuable glossary; and a comprehensive reference list. The US Center for Disease Control and US Environmental Protection Agency also maintain extensive databases on lead.

It is not readily apparent in most general audience publications whether lead compounds are also of concern (in addition to metallic lead), and whether they are regulated. An extensive review of the scientific literature, government regulations, and toxicology databases indicates, however, that within these circles, lead and lead compounds (inorganic) are grouped together and treated equally. At issue, actually, is the bioavailability of the lead, regardless of its source. Both lead and lead compounds are generally introduced into the body by either inhalation or ingestion. Inhalation is a greater concern because inhaled lead is almost entirely absorbed by the body, while only a fraction of ingested lead is absorbed [38]. The primary sink for inorganic lead is the blood. Within the body, lead is transported as if it were calcium, i.e., to soft tissues, teeth, and bones. Human exposure to inorganic lead and its compounds can result from various sources: occupational exposure, general exposure, environmental exposure (which affects general exposure), and hazardous waste disposal.

From the occupational exposure perspective, both the lead oxides used in ceramics and glasses and the lead-containing minerals from which lead is derived present risks, although appropriate precautions are generally taken to protect workers. Examples of occupational exposure warnings and carcinogen levels are presented in Table 9. In addition to these warnings about human exposure, inorganic lead and lead compounds also present a variety of dangers through chemical reactions either with other substances or upon heating [37,39].

From the human health (general exposure) perspective, elemental lead and inorganic lead compounds are identified as possible human carcinogens by the International Agency for Research on Cancer (IARC); they are also listed in the Hazardous

Table 9 Carcinogen levels and occupational exposure warnings for lead and various lead compounds [39]

Substance	CAS No.[a]	Carcinogen level[b]	Warnings[c]	Reference
Lead	7439-92-1	A3	Dust, women	ICSC 0052
PbO	1317-36-8	A3	Dust, women	ICSC 0288
PbO_2	1309-60-0	A3	Women	ICSC 1001
Pb_3O_4	1314-41-6	A3	Women	ICSC 1002
$PbCO_3$	598-63-0	A3	All contact	ICSC 0999
$PbCrO_4$	7758-97-6	A2	All contact, dust, women, children	ICSC 0003

[a] CAS: Chemical Abstracts Services
[b] A3: Animal carcinogen; A2: Suspected human carcinogen
[c] Dust, prevent dispersion of dust; women, avoid exposure of (pregnant) women; all contact, avoid all contact; children, avoid exposure of adolescents and children

Substances Data Bank provided by the National Library of Medicine. The US Environmental Protection Agency has classified lead as a Group B2 (probable) carcinogen and as a Category I contaminant (which results in the Maximum Contaminant Level Goal (MCLG) for drinking water being set at a value of zero) [40]. In response to EPA's classification, the State of California now regulates lead and lead compounds through Proposition 65, which requires labeling of all products that contain cancer-causing agents. The US Federal Drug Administration (FDA) also monitors lead exposure, because fine dust can settle on food or lead can leach from lead-containing food containers. Lead and lead compounds are regulated with respect to air quality standards, drinking water standards, and blood levels. Examples of some of these standards are presented in Table 10.

From the environmental loading perspective, lead and lead compounds are regulated by EPCRA (the Emergency Planning and Community Right To Know Act, 1986) as a persistent, bioaccumulative, and toxic (PBT) chemical. As such the disposal and release of these substances are subject to Toxic Release Inventory (TRI) reporting. In 2001, the threshold reporting level for lead and lead compounds was lowered from 5,000 to 220 kg per year. Results from the 2002 TRI Report [42] indicate that in the United States over 440,000 tons of lead and lead compounds were disposed of or released into the environment. This represents more than 97% of all PBT chemical releases in that year. Lead and lead compounds are also regulated by RCRA (Resource Conservation and Recovery Act, 1976). Thus, products that contain lead or lead compounds must be treated as hazardous waste. Discards of lead from glass and ceramic products into the municipal solid waste stream have been increasing, primarily as the result of CRT disposal. The US EPA estimated that lead discards from TV glass increased from 10,000 tons in 1970 to over 52,000 tons in 1986; lead discards from light bulb glass increased from ~500 tons to almost 700 tons in the same period; and for other glass and ceramic applications combined, the increase was from 3,300 tons to almost 7,800 tons [43].

In the 1970s, a major concern was the documented evidence that the lead in lead-containing glazes used on whitewares used as food containers and for cooking could leach lead into food. The glass industry was responsive to these and related occupational concerns, established appropriate operating procedures and monitoring systems, as well as reduced the use of raw materials that were more soluble, such as lead carbonates [14,15,31,44]. Later, leaching of lead from leaded crystal, especially that used for wine decanters, became a concern [45]. This concern still exists, but is mitigated

Table 10 Representative regulatory limits and guidelines for lead (in metallic lead and various inorganic lead compounds) in selected fates

Guideline/regulation	Organization	Limit	Units	Fate	G/R	Reference
Air content	US EPA	0.0015	mg m^{-3}	Air	R	36
Permissible exposure limit (TWA)	US OSHA	0.0500	mg m^{-3}	Air	R	41
Recommended exposure limit (TWA)	NIOSH	0.0500	mg m^{-3}	Air	R	41
Threshold limit value (TWA)	ACGIH	0.0500	mg m^{-3}	Air	G	39
Blood lead level of concern in children	US CDC	0.0100	mg dL^{-1}	Blood	G	37
Blood lead level of concern	WHO	0.0200	mg dL^{-1}	Blood	G	37
Blood lead level of concern	ACGIH	0.0300	mg dL^{-1}	Blood	G	37
Blood lead level of concern	US OSHA	0.0400	mg dL^{-1}	Blood	G	37
Blood lead level – medical treatment in children	US CDC	0.0450	mg dL^{-1}	Blood	G	36
Blood lead level – medical removal	US OSHA	0.0500	mg dL^{-1}	Blood	G	37
Leaching solution for cups and mugs	US FDA	0.0050	mg mL^{-1}	Food	R	37
Leaching solution for pitchers	US FDA	0.0050	mg mL^{-1}	Food	R	37
Leaching solution for ceramicware flatware	US FDA	0.0300	mg mL^{-1}	Food	R	37
Maximum contaminant level	US EPA	0.0500	mg L^{-1}	Landfill	R	37
Toxicity characteristic leaching protocol limit	US EPA	0.1500	mg L^{-1}	Landfill	R	37
Drinking water action level	US EPA	0.0150	mg L^{-1}	Water	R	37
Maximum contaminant level goals	US EPA	0.0000	mg L^{-1}	Water	G	37
Drinking water guidelines	WHO	0.0500	mg L^{-1}	Water	G	37

G/R: G = Guideline, R = Regulation; *TWA*: time weighted average; *US EPA*: US Environmental Protection Agency; *US OSHA*: US Department of Occupational Safety and Health Administration; *NIOSH*: National Institute for Occupational Safety and Health; *ACGIH*: American Conference of Governmental Industrial Hygienists; *US CDC*: *US* Center for Disease Control; *WHO*: World Health Organization; *US FDA*: US Food and Drug Administration

somewhat through educational programs such as those required by Proposition 65. At present, a major concern is the proliferation of electronic waste, which includes cathode ray tubes [38]. Recent studies document concern for leaching of lead from CRTs when exposed to simulated landfill conditions, i.e., using the Toxicity Characteristic Leaching Procedure (TCLP) test protocol established by the US EPA [46].

As a consequence of these and other documented concerns over lead poisoning, various new legislative issues have come into place that will further limit the future use of lead. In the European Union, for instance, the Reduction of Hazardous Substances (RoHS) Directive has forced the removal of lead from electronics, worldwide; in the United States, the States of California, Massachusetts, Maine, and Minnesota have banned the disposal of CRTs in landfills, forcing special handling and encouraging recycling; in Japan, lead-free products have been embraced and utilized as a marketing tool. These new laws and marketing pressures, plus the plethora of existing rules and regulations, are forcing industry to consider alternative materials that do not contain lead. Examples include lead-free glasses for lamp applications [43,47], for cathode ray tubes [43,48,49], and for glazes [43,49], as well as lead-free oxides to replace PZTs and PLZTs [43,50].

5 Summary

This chapter has provided an overview of lead and lead compounds, as used in glass and ceramic products. Some of the basic physical characteristics of (metallic) lead, lead-containing minerals such as galena (PbS), and lead oxides have been provided. The lead oxide most important to glass and ceramic fabrication is litharge, PbO. Lead (oxide) is used in glasses for several reasons: to increase the refractive index of the glass, to decrease the viscosity of the glass, to increase the electrical resistivity of the glass, and to increase the X-ray absorption capability of the glass. For ceramic applications, which are primarily ferroelectric applications, the main reason to include lead (oxide) in the material is because it can significantly increase the Curie point. Leaded glasses have a wide range of chemical composition, from 2 to 77% PbO by weight, depending on the application; lead-containing ceramics typically contain 55–70 wt% Pb.

The lead oxides used in glasses and ceramics are derived from both primary lead-containing minerals and secondary recycled leaded glass. Several process steps are required to mine, concentrate, extract, smelt, and refine the lead, which is then oxidized to form lead oxide. Lead (and inorganic lead compounds such as lead oxides) is known to be toxic and probably carcinogenic to humans. As a result, these substances are highly regulated and the potential for lead exposure through water, land, and air is closely monitored. The legislative burden on lead users continues to increase, which has led to significant efforts to find lead-free alternatives for both glass and ceramic products.

Acknowledgments I would like to acknowledge the research assistance of Xiaoying Zhou and Tammy Tamayo, as well as Valerie Thomas and Dele Ogunseitan for their guidance on the health risks associated with lead.

References

1. M.F. Ashby, *Materials Selection in Mechanical Design,* 3rd edn., Elsevier, San Francisco, 2005. ISBN 0-7506-6168-2.
2. Amethyst Galleries' Mineral Gallery, Amethyst Galleries, Inc., St. Augustine, FL, 2004.
3. *Standard Thermodynamic Properties of Chemical Substances,* CRC Press LLC, 2000.
4. G.W. McLellan and E.B. Shand, *Glass Engineering Handbook,* 3rd edn., McGraw-Hill, San Francisco, 1984.
5. Y.-M. Chiang, D.P. Birnie, III, and W.D. Kingery, *Physical Ceramics: Principles for Ceramic Science and Engineering,* Wiley, New York, 1997.
6. R.A. Flinn and P.K. Trojan, *Engineering Materials and Their Applications,* 4th edn., Wiley, New York, 1994.
7. W.F. Smith, *Foundations of Materials Science and Engineering,* 3rd edn., McGraw-Hill, San Francisco, 2004.
8. M. Barsoum, *Fundamentals of Ceramics,* McGraw-Hill, San Francisco, 1997.
9. W.D. Kingery, H.K. Bowen, and D.R. Uhlmann, *Introduction to Ceramics,* 2nd edn., Wiley, New York, 1976.
10. M.J. Matthewson, Design properties for glass and glass fibers, in *Engineered Materials Handbook,* Vol. 4: *Ceramics and Glasses,* ASM International, 1991, pp. 741–745.
11. G.J. Fine, Consumer houseware applications, in *Engineered Materials Handbook,* Vol. 4: *Ceramics and Glasses,* ASM International, 1991, pp. 1100–1103.

12. R.A. Eppler and L.D. Gill, Glazes and Enamels, in *Engineered Materials Handbook*, Vol. 4: *Ceramics and Glasses*, ASM International, 1991, pp. 1061–1068.

13. R.A. Eppler, Glazes and Enamels, in *Advances in Ceramics*, Vol. 18: *Commercial Glasses*, D.C. Boyd and J.F. MacDowell, (eds.), American Ceramic Society, Westerville, OH, 1986, pp. 65–78.

14. J.E. Marquis, Lead in glazes – Benefits and Safety Precautions, *Am. Ceram. Soc. Bull.*, **50**(11), 921–923 (1971).

15. R.A. Eppler, Formulation of glazes for low Pb release, *Am. Ceram. Soc. Bull.*, **54**(5), 496–499 (1975).

16. E.W. Deeg, Optical glasses, in *Advances in Ceramics*, Vol. 18: *Commercial Glasses*, D.C. Boyd and J.F. MacDowell (eds.), American Ceramic Society, Westerville, OH, 1986, pp. 9–34.

17. E.W. Deeg, Ophthalmic and Optical Glasses, in *Engineered Materials Handbook*, Vol. 4: *Ceramics and Glasses*, ASM International, 1991, pp. 1074–1081.

18. J.H. Connelly and D.J. Lopata, CRTs and TV picture tubes, in *Engineered Materials Handbook*, Vol. 4: *Ceramics and Glasses*, ASM International, 1991, pp. 1038–1044.

19. J.R. Coaton and A.M. Marsden, *Lamps and Lighting*, 4th edn., Arnold/Wiley, London/ New York, 1997.

20. N. Braithwaite and G. Weaver, *Electronic Materials: Inside Electronic Devices (Materials in Action Series)*, 2nd edn., The Open University, Boston, 2000.

21. N.P. Bansal and R.H. Doremus, *Handbook of Glass Properties*, Academic Press, San Diego, 1986.

22. P.R. Prud'homme van Reine, W.J. van den Hoek, and A.G. Jack, Lighting, in *Engineered Materials Handbook*, Vol. 4: *Ceramics and Glasses*, ASM International, 1991, pp. 1032–1037.

23. R.R. Tummala and R.R. Shaw, Glasses in Microelectronics in the information-processing industry, in *Advances in Ceramics*, Vol. 18: *Commercial Glasses*, D.C. Boyd and J.F. MacDowell (eds.), American Ceramic Society, Westerville, OH, 1986, pp. 87–102.

24. L.L. Hench and J.K. West, *Principles of Electronic Ceramics*, Wiley, New York, 1990.

25. S.O. Kasop, *Principles of Electronic Materials and Devices*, 2nd edn., McGraw Hill, San Francisco, CA, 2002.

26. I. Burn, Ceramic capacitor dielectrics, in *Engineered Materials Handbook*, Vol. 4: *Ceramics and Glasses*, ASM International, 1991, pp. 1112–1118.

27. K. Uchino, Piezoelectric ceramics, in *Engineered Materials Handbook*, Vol. 4: *Ceramics and Glasses*, ASM International, 1991, pp. 1119–1123.

28. G.H. Haertling, Electrooptic ceramics and devices, in *Engineered Materials Handbook*, Vol. 4: *Ceramics and Glasses*, ASM International, 1991, pp. 1124–1130.

29. G.R. Smith, Lead, *Minerals Yearbook*, US Geological Survey, 2002.

30. M. King and V. Ramachandran, Lead, *Kirk-Othmer Encyclopedia of Chemical Technology*, Wiley, New York, 1995.

31. Lead Processing, *Encyclopedia Britannica*, 2003.

32. *Emission Estimation Technique Manual for Lead Concentrating, Smelting and Refining*, National Pollutant Inventory, Environment Australia, 1999.

33. Current Metal and Scrap Prices, London Metal Exchange (LME), August 12, 2004.

34. P.N. Gabby and J.I. Martinez, Lead in April 2004, *Mineral Industry Surveys*, US Geological Survey, July 2004.

35. 2003 CERCLA Priority List of Hazardous Substances, Agency for Toxic Substances and Disease Registry, Atlanta, GA, May 2004.

36. Lead, CAS # 7439-92-1, Agency for Toxic Substances and Disease Registry ToxFAQs, ATSDR, Atlanta, GA, June 1999.

37. Toxicological Profile for Lead, US Department of Health and Human Services, Public Health Service, Agency for Toxic Substances and Disease Registry, Atlanta, GA, July 1999.

38. J.M. Schoenung, O.A. Ogunseitan, J.-D. M. Saphores, and A.A. Shapiro, Adoption of Pb-free electronics: policy differences and knowledge gaps, *J. Ind. Ecol.*, **8**(4) (2004).

39. International Labour Organization, International Occupational Safety and Health Information Centre (CIS): Lead, ICSC 0052, 2002; Lead (II) Oxide, ICSC 0288, 2003; Lead Dioxide, ICSC 1001, 2002; Lead Tetroxide, ICSC 1002, 2002; Lead Carbonate, ICSC 0999, 2001; Lead Chromate, ICSC 0003, 2003, Geneva, Switzerland.

40. Maximum contaminant level goals and national primary drinking water regulations for lead and copper, *Fed. Regist.*, **56**(110) 26470 (1991).

41. NIOSH Pocket Guide to Chemical Hazards: Lead, National Institute for Occupational Safety and Health, Washington, D.C., 2004.
42. TRI Disposal or Other Releases, PBT Chemicals, 2002, US Environmental Protection Agency, April 2004.
43. *Preliminary Use and Substitutes Analysis of Lead and Cadmium in Municipal Solid Waste, EPA 530-R-92-010*, US Environmental Protection Agency, April 1992.
44. J.S. Nordyke, Lead products, *Am. Ceram. Soc. Bull.*, **65**(5) 737–738 (1986).
45. J.H. Graziano and C. Blum, Lead exposure from lead crystal, *The Lancet*, **337**, 141–142 (1991).
46. T.G. Townsend, S. Mussen, Y.-C. Jang, and I.-H Chung, Characterization of lead leachability from cathode ray tubes using the toxicity characteristic leaching procedure, Florida Center for Solid and Hazardous Waste Management, Report 99-5, Gainesville, FL, 1999.
47. B. Filmer, European Patent No. 0 603 993 A1, 1994.
48. P. Hedemalm, Some uses of lead and their possible substitutes, KEMI Report No. 3/94, The Swedish National Chemicals Inspectorate, PrintGraf, Stockholm, Sweden, Feb. 1994.
49. *Lead Review*, Nordic Council of Ministers, Copenhagen, Norway, January 2003.
50. I. Campbell, Lead and cadmium free glasses and frits, *Glass Technol.*, **39**(2) 38–41 (1998).
51. M. Ichiki, L. Zhang, M. Tanaka, and R. Maeda, Electrical properties of piezoelectric sodium-potassium niobate, *J. Eur. Ceram. Soc.*, **24** 1693–1697 (2004).

Chapter 10
Zirconia

Olivia A. Graeve

Abstract Zirconia is a very important industrial ceramic for structural applications because of its high toughness, which has proven to be superior to other
ceramics. In addition, it has applications making use of its high ionic conductivity.
The thermodynamically stable, room temperature form of zirconia is baddeleyite.
However, this mineral is not used for the great majority of industrial applications of
zirconia. The intermediate-temperature phase of zirconia, which has a tetragonal structure, can be stabilized at room temperature by the addition of modest amounts (below
~8 mol%) of dopants such as Y^{3+} and Ca^{2+}. This doped zirconia has mechanical toughness values as high as $17 MPa \cdot m^{1/2}$. On the other hand, the high-temperature phase
of zirconia, which has a cubic structure, can be stabilized at room temperature by the
addition of significant amounts (above ~8 mol%) of dopants. This form of zirconia has
one of the highest ionic conductivity values associated with ceramics, allowing the use
of the material in oxygen sensors and solid-oxide fuel cells. Research on this
material actively continues and many improvements can be expected in the years to
come.

1 Introduction

Zirconia (ZrO_2) is an extremely versatile ceramic that has found use in oxygen pumps
and sensors, fuel cells, thermal barrier coatings, and other high-temperature applications, all of which make use of the electrical, thermal, and mechanical properties of
this material. Proof of the interest and usefulness of zirconia can be seen from the
voluminous literature found on this material. This chapter is intended to provide a
concise summary of the physical and chemical properties of all phases of zirconia that
underlie the appropriate engineering applications.

The three low-pressure phases of zirconia are the monoclinic, tetragonal, and
cubic, which are stable at increasingly higher temperatures. Calculated energy vs.
volume data at zero absolute temperature confirms the higher stability of the monoclinic
phase (Fig. 1). However, most engineering applications make use of the tetragonal and
cubic phases, even though their stability at low temperatures is quite low. In fact,
the engineering use of all three phases of zirconia in pure form is rare. Generally,

J.F. Shackelford and R.H. Doremus (eds.), *Ceramic and Glass Materials:*
Structure, Properties and Processing.
© Springer 2008

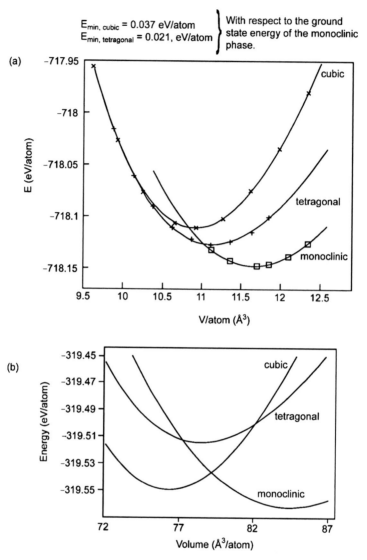

$E_{min, cubic}$ = 0.037 eV/atom
$E_{min, tetragonal}$ = 0.021, eV/atom

With respect to the ground state energy of the monoclinic phase.

Fig. 1 Computed energy vs. volume data for cubic, tetragonal, and monoclinic phases from (**a**) Stapper et al. [1] and (**b**) Dewhurst and Lowther [2] (reprinted with permission)

zirconia is doped with oxides such as Y_2O_3 that stabilize the high-temperature phases at room temperature. This has enormous consequences for both the mechanical and electrical properties of zirconia, even though the local atomic and electronic structure of Zr^{4+} in all three polymorphs is for the most part dopant independent [3].

Doping of zirconia results in stabilization of the tetragonal phase at lower dopant concentrations (for mechanical toughness) or the cubic phase at higher dopant concentrations (for high ionic conductivity) at room temperature. The stabilization of the tetragonal phase at room temperature can result in the following common forms of zirconia: (1) partially stabilized zirconia (PSZ) – zirconia consisting of a matrix of a brittle ceramic and a dispersion of tetragonal precipitates, where the tetragonal precipitates can either be in pure form or doped with Ca^{2+} (Ca-PSZ) or Mg^{2+} (Mg-PSZ); and (2) tetragonal zirconia polycrystals (TZP) – zirconia consisting of a matrix of stabilized ZrO_2 that has been stabilized in the tetragonal form by the addition of dopants such as Ce^{4+} (Ce-TZP) and Y^{3+} (Y-TZP). Fully-stabilized zirconia (FSZ) refers to a material that has been completely stabilized in the cubic form.

Stabilized zirconia in thermal barrier coatings (TCB) is ubiquitous, finding itself in combustor liners, transition sections, nozzle guide vanes, and rotor blades. It is one of the most used ceramics for TCB applications because of its low thermal conductivity, high-temperature stability in oxidizing and reducing environments, coefficient of thermal expansion similar to iron alloys, high toughness, and cost-effectiveness by which it can be applied onto metal surfaces. Its use allows a 200°C increase in the operational temperature of the engine, resulting in a much higher efficiency [4].

The second, well-known use of stabilized zirconia is in oxygen sensors. These types of devices make use of the very high ionic conductivity of Y_2O_3- or CaO-doped cubic zirconia. The sensor assembly consists of a zirconia tube with one end closed. The inside of this tube is exposed to air and the outside is exposed to the gas that requires measurement of oxygen levels. When there is a difference in oxygen partial pressure between the inside and outside, oxygen is transported across the ceramic tube. This transport results in a measurable voltage.

A solid-oxide fuel cell (SOFC) functions similar to an oxygen sensor. An SOFC converts the chemical energy of a fuel directly to electrical energy and heat and consists of two electrodes that sandwich an electrolyte, allowing ions to pass while blocking electrons. The air electrode allows oxygen to pass through to the electrolyte. At the electrolyte interface, the oxygen dissociates into ions that travel across the electrolyte via ionic conduction. Typical SOFC's consist of an Y_2O_3-doped ZrO_2, with about 8 mol% yttrium, as the electrolyte. At the fuel electrode, the oxygen ions that have traveled across the electrolyte react with the fuel forming H_2O and possibly other gases, depending on the type of fuel used. During the reaction, at the fuel electrode/electrolyte interface, electrons are generated that travel through an external circuit, thus generating electrical current that can be used for doing external work. This technology will become increasingly important as a "clean" source of electricity as pressures on the environment from the use of coal and petroleum continue to increase.

2 Crystalline and Noncrystalline Structures

The complex crystallography of zirconia plays an important role in the challenges to develop commercially viable applications of this material. At room temperature, bonding in this material is a combination of ionic and covalent and results in a structure in which zirconium is seven-coordinated, which is rather unusual and is a product of the large difference in ionic sizes between zirconium and oxygen. The formation of this material from pure α-Zr and oxygen starts at around 23 at.% O, corresponding to a composition of $ZrO_{0.3}$ [5]. The perfect stoichiometry for this material in which there is one zirconium and two oxygen ions for each formula unit is not used in industrial applications. Doping of the structure produces oxygen vacancies resulting in chemical formulas of the type $ZrCa_xO_{2-x}$, for the case of CaO doping, to maintain charge balance in the structure.

2.1 Cubic Zirconia

Cubic zirconia (Fm3m) has a fluorite structure with one formula unit ($Z = 1$) in the primitive cell. This cell contains one zirconium ion located at (0, 0, 0) and coordinated with eight equidistant oxygen ions. The two oxygen ions are located at (¼, ¼, ¼) and (¾, ¾, ¾), both tetrahedrally coordinated to four zirconium ions. The nonprimitive face-centered arrangement is illustrated in Fig. 2 and contains four zirconium ions located at (0, 0, 0), (½, ½, 0), (½, 0, ½), and (0, ½, ½). The eight oxygen ions are located at (¼, ¼, ¼), (¼,¼, ¾), (¼, ¾, ¼), (¾, ¼, ¼), (¾, ¾, ¼), (¾, ¼, ¾), (¼, ¾, ¾), and (¾, ¾, ¾). The translational vectors for this structure are

$$t_1 = \left(0, \tfrac{a}{2}, \tfrac{a}{2}\right),$$
$$t_2 = \left(\tfrac{a}{2}, 0, \tfrac{a}{2}\right), \tag{1}$$
$$t_3 = \left(\tfrac{a}{2}, \tfrac{a}{2}, 0\right).$$

General crystallographic correlations for this structure are given in Table 1. Experimental measurements of the lattice parameter, a_c, at 2,683, 2,388, and 2,503 K result in 0.5269, 0.52438, and 0.5247 nm, respectively. The measurement at 2,683 K was done

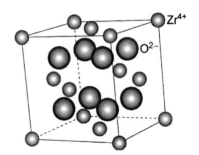

Fig. 2 Crystal structure of cubic zirconia as seen from the [153] direction

Table 1 General crystallographic correlations in cubic zirconia

Space group	Fm3m
Reflection conditions	hkl: $h + k$, $k + l$, $l + h = 2n$
	$0kl$: k, $l = 2n$
	hhl: $h + l = 2n$
	$h00$: $h = 2$
Coordination number	8
Z	1
Lattice parameter	a_c
Unit cell volume	$V_c = a_c^3$
Ionic positions	Zr^{4+}: 4a
	O^{2-}: 8c

in a neutral atmosphere, while the latter two measurements were done in a reducing atmosphere [6]. At increasing pressures, the lattice parameter changes to 0.4947, 0.4925, and 0.4916 nm for pressures of 28.9, 33.8, and 37.3 GPa, respectively [7].

2.2 Tetragonal Zirconia

The tetragonal zirconia structure (Fig. 3 [6], Table 2), with space group $P4_2/nmc$ (primitive) [8, 9] and a cation coordination number of 8, is derived from the cubic fluorite structure by the movement of oxygen anions along one of the cubic axes, which results in a tetragonal distortion along that axis, as shown in Fig. 3 for a distortion along the c-axis. The two zirconium ions in the primitive structure are located at (0, 0, 0) and (½, ½, ½), and the four oxygen ions are located at (0, ½, z), (½, 0, −z), (0, ½, ½ + z), and (½, 0, ½ −z), where z = 0.185. This results in a body-centered tetragonal (bct) structure, which is sometimes described as a pseudofluorite structure.

The transition from cubic to tetragonal is displacive in which four Zr^{4+} cations in the 4a cubic positions split into two groups to occupy the 2b positions in the tetragonal structure, and the O^{2-} anions in the 8c cubic positions also split into two groups to occupy the 4d positions in the tetragonal structure. The directions of a and b axes in the tetragonal primitive lattice are 45° from those in the cubic cell. The c axis in both structures remains the same. Following the literature convention of reporting the parameter $d_z = 0.25 − z$, where z represents the third coordinate of the oxygen position 4d (0, ½, z), the calculated c/a ratio and internal parameter d_z of the tetragonal phase as a function of volume are shown in Fig. 4 [2].

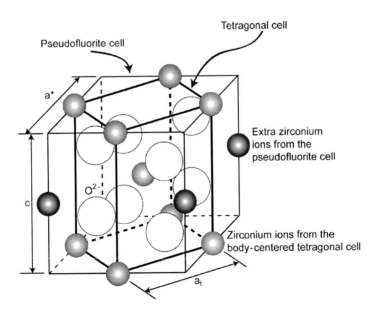

Fig. 3 Tetragonal zirconia unit cell in both the body-centered tetragonal and pseudofluorite descriptions (adapted from Aldebert and Traverse [6])

Table 2 General crystallographic correlations in tetragonal zirconia

Space group	$P4_2/nmc$
Reflection conditions	$hk0: h + k = 2n$
	$hhl: l = 2n$
	$00l: l = 2n$
	$h00: h = 2n$
Coordination number	8
Z	2
Lattice parameters	$a_t = b_t \approx a^*/\sqrt{2}$
	$c_t \approx a^*$
Unit cell volume	$V_t \approx V_c/2$
Ionic positions	$Zr^{4+}: 2b$
	$O^{2-}: 4d$

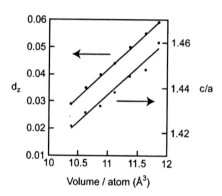

Fig. 4 Calculated c/a ratio and internal structural parameter, d_z, of the tetragonal phase as a function of volume [2] (reprinted with permission)

2.3 Monoclinic Zirconia

EXAFS analysis by Winterer [10] and Li et al. [11] shows that monoclinic ZrO_2 can be described by a sevenfold Zr–O shell with an average distance of 0.216 nm and a mean vibrational frequency of 410 cm^{-1}. The structure consists of layers of triangular coordination polyhedra of three O_1–Zr bonds and four distorted tetrahedral O_2–Zr bonds for a total of seven oxygen ions surrounding the zirconium [10], as shown in Fig. 5. In this illustration, there are three unit cells that help with visualization. The Zr^{4+} sublattice is marked with dashed lines and the seven coordinating oxygen ions surrounding the darker-colored zirconium ion are marked with numbers. The plane that constitutes the O_1 atoms is nearly parallel to the plane of O_2 atoms. The general crystallographic correlations for this structure are listed in Table 3. The Zr–O distances range from 0.1885 to 0.2360 nm for the O_1–Zr shell and 0.1914 to 0.2511 nm for the O_2–Zr shell.

There are also three distinct Zr–Zr subshells at 0.346,[1] 0.396 (0.401),[2] and 0.454 (0.455) nm with a total of 12 zirconium next nearest neighbors at an average distance of 0.372 nm. The coordination numbers for the first, second, and third subshells of the

---

[1] Winterer and Li et al. report the same value.

[2] First value by Winterer; value in parentheses by Li et al.

Fig. 5 Crystal structure of monoclinic zirconia as seen from the [1$\bar{3}$1] direction

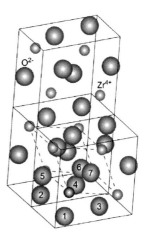

Table 3 General crystallographic correlations in monoclinic zirconia

Space group	$P2_1/c$
Reflection conditions	$h0l: h + l = 2n$
	$0k0: k = 2n$
	$h00: h = 2n$
	$00l: l = 2n$
Coordination number	7
Z	4
Lattice parameters	$a_m \neq b_m \approx a_c$
	$c_m > a_c$
Unit cell volume	$V_m \approx 2V_c$
Ionic positions	$Zr^{4+}: 4e$
	$O_I^{2-}: 4e$
	$O_{II}^{2-}: 4e$

Zr–Zr bonds are 7, 4, and 1, respectively. The zirconium ions form layers parallel to the (100) planes of the unit cell, with the O_1 ions on one side and the O_2 ions on the other side. The distance between two layers of zirconium ions is larger when they are separated by O_1 ions.

2.4 High-Pressure Phases

High-pressure experimental measurements on ZrO_2 have revealed that the ambient monoclinic baddeleyite phase transforms under increasingly higher pressures to a series of orthorhombic phases. The first orthorhombic phase starts appearing at an applied pressure of about 3.5 GPa [13], depending upon the crystallite size of the material – lower crystallite size results in a higher transformation pressure [14], although the phase transformation is not completed until 10 ± 1 GPa, as determined by Desgreniers and Lagarec [15]. The calculated (ab initio) transition pressure, according to Stapper et al. [2], is 5.7 GPa, so experimental measurements and ab initio calculations are in general agreement. This first orthorhombic phase (Pbca [16], $Z = 8$, coordination number = 7) is observed to exist up to about 25 GPa when a second orthorhombic structure appears (Pnma, $Z = 4$, coordination number = 9), although the precise onset pressure has not been determined accurately. This phase is stable at ambient temperature up to at least 70 GPa. A projection of the crystal structure is illustrated in Fig. 6 [17]. The change in volume with pressure is shown in Fig. 7 [15], where the initial volume, V_o, is taken as 70.32 Å.

Aside from the two high-pressure phases of zirconia, a hexagonal high-temperature and high-pressure phase was found by Ohtaka et al. [18] by quenching pure ZrO_2 powders from above 1,000°C and 20 GPa. This hexagonal structure ($Z = 8$) reverts to the baddeleyite structure when pressure is released below 1 GPa.

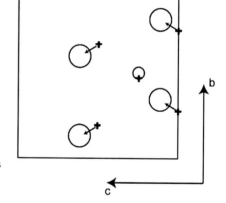

Fig. 6 A projection into (100) of the orthorhombic-I structure. The *crosses* indicate the atom positions in the tetragonal structure, and the *arrows* the presumed displacements of these atoms during the transition to orthorhombic [17] (reprinted with permission)

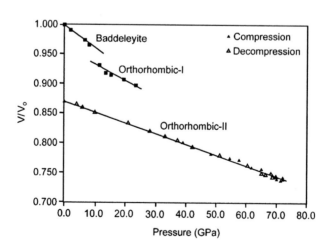

Fig. 7 Pressure dependence of the volume of the dense phases of zirconia [15] (reprinted with permission)

2.5 Amorphous Structure

When ZrO_2 powders are prepared using any number of precipitation methodologies, the resulting structure before calcination is amorphous [19–22]. Detailed studies by Chadwick et al. [23] have shown that the amorphous gel has both OZr_3 and OZr_4 environments in approximately equal proportions, as evidenced from the two ^{17}O NMR peaks at 405 ppm and 303 ppm, respectively. EXAFS results show that there is a well-defined oxygen shell with a coordination number of 7 at ~0.214 nm and a second shell with a much smaller coordination number at ~0.342 nm corresponding to the Zr–Zr correlation. The NMR and Zr K-edge EXAFS results unambiguously indicate that the short-range structure of the amorphous gel is monoclinic-like.

Once the amorphous gel is heated, ^{17}O NMR shows that both OZr_3 and OZr_4 environments remain, although there is an increase in line width. There is also an increase in the isotropic chemical shift of the peaks, especially the OZr_4 peak, which moves from 303 to 321 ppm as the gel starts to crystallize. At the crystallization temperature (approximately 360°C), EXAFS results show that there is a distinct change in the structure, with the oxygen correlation now better fit by two closely spaced shells and a large increase in the coordination number to 12 associated with the Zr–Zr correlation. This is likely because the particles are nanocrystalline.

After crystallization has occurred, there is an additional ^{17}O NMR peak at 374 ppm that corresponds to OZr_4 in tetragonal ZrO_2. However, the NMR data show that although crystalline tetragonal ZrO_2 is forming at the point of crystallization, oxygen is still present as part of the disordered ZrO_2. Upon further heating, the tetragonal content increases significantly, but at no time is there complete elimination of the disordered ZrO_2.

In addition, ^1H NMR results show that, even after crystallization, it is not correct to describe the sample composition as ZrO_2. Data show that, after being heated to 300°C, the sample's composition is $ZrO_{1.42}(OH)_{1.16}$. At 500°C, well above the crystallization temperature, it is still $ZrO_{1.76}(OH)_{0.48}$. The OH^- content found below this temperature is not related to the usual surface hydroxylation upon exposure to the atmosphere, instead the hydroxyls are structural units within the sample. Above 700°C, the hydroxide content is no longer measurable and with subsequent heating the sample changes to the monoclinic structure. Hence, the reaction for the formation of zirconia by precipitation can be described as:

$$Zr_4O_{(8-x)}(OH)_{2x} \rightarrow ZrO_{(2-y/2)}(OH)^y (amorphous) \rightarrow$$

$$ZrO_{(2-z/2)}(OH)_z (tetragonal, crystalline; monoclinic - like, disordered) \rightarrow$$

$$ZrO_2 (monoclinic, crystalline)$$

There is an initial metal hydroxide that becomes an amorphous oxide containing hydroxyls. With heating, some of these hydroxyls are lost, resulting in the formation of a mixture with more-ordered tetragonal and less-ordered monoclinic components. With further heating, the eventual crystalline product becomes monoclinic.

In contrast, an EXAFS analysis from amorphous zirconia films of nominal ZrO_2 composition, as opposed to a hydroxide composition, found that the local structure in amorphous ZrO_2 can be described by an eightfold Zr–O shell widely spread between 0.19 and 0.32 nm with a distinct peak at 0.216 nm consisting of four oxygen nearest neighbors. The average Zr–O coordination distance is 0.255 nm for all eight oxygen neighbors. The local structure also consists of a very broad Zr–Zr shell at about 0.41 nm with 12 next nearest neighbors [10].

3 Point Defects

Point defects play a central role in the use of zirconia ceramics in such applications as oxygen sensors and fuel cells. As a result, point defects in these materials have been extensively studied.

3.1 Interstitial Defects

Interstitial defects in monoclinic zirconia have been modeled in detail by Foster et al. [24]. Using plane wave density functional theory, the tetragonal bonding and triple-planar bonding geometries of lattice oxygen ions were determined. In addition, it was determined that interstitial defects can form stable defect pairs with either type of lattice oxygen ions (i.e., tetragonal or triply bonded). The analysis looked at defect pairs formed by interstitial oxygen ions with three possible charge states: 0, −1, and −2, bonded to triple-planar lattice oxygen ions. An analysis of oxygen vacancies both in the triple-planar and tetragonal geometries was also undertaken.

A neutral oxygen interstitial forming a defect pair with a triple-bonded oxygen is illustrated in Fig. 8 [24]. Using the oxygen atomic energy as a reference, a single neutral oxygen can be incorporated in the lattice as an interstitial with an energy gain of −1.6 eV, if next to a triple-bonded lattice oxygen, and −0.8 eV, if next to a tetragonally bonded lattice oxygen. Figure 8 illustrates the fully relaxed charge density and positions of ions, showing that the interstitial and lattice oxygen form a strong covalent bond. The labels A and B associated with the lattice ions represent two different crystal planes within the structure. The lattice oxygen (O_A), forming the defect pair with the interstitial oxygen, relaxes by up to 0.05 nm to accommodate the interstitial, distorting the triply-bonded oxygen with respect to the three zirconium ions bonded to it. The O–3Zr group has a slight pyramidal shape with its apex pointing away from the interstitial. The rest of the crystal remains more or less undisturbed, with the nearest zirconium (Zr_A) only relaxing by about 0.005 nm. The case of a singly-charged oxygen interstitial forming a defect pair with a triply-bonded oxygen results in weakening of the covalent bond between the defect pair significantly. The extreme is the case of a doubly-charged oxygen interstitial in which the interstitial forms elongated bonds with the zirconium ions and occupies a new triple site, which is bonding with the Zr_A, Zr_B, and a new zirconium ion.

3.2 Vacancy Defects

Oxygen vacancies in cubic zirconia result in a calculated displacement pattern as shown in Fig. 9 [2]. In this figure, the vacancy is depicted as a small cube, the oxygen

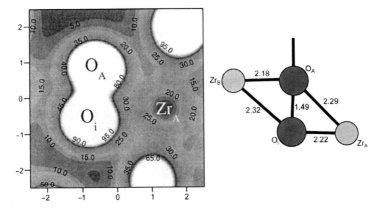

Fig. 8 Charge density in the plane through Zr_A, O_A, and O_i, and a schematic diagram of a neutral oxygen interstitial (O_i) near a triple-bonded oxygen (O_A) in zirconia. Charge density is in $0.1\,eV\,\text{Å}^{-1}$ and all distances are in Å [24] (reprinted with permission)

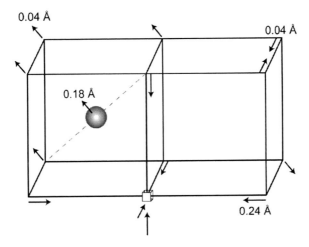

Fig. 9 Displacement pattern of atoms around an isolated vacancy in a 95 atom supercell [2] (reprinted with permission)

atoms occupy the sites at the corners of the cubes, and metal cations occupy half of the sites at the center of the cubes. The six oxygen neighbors nearest to the vacancy move along <100> by 0.024 nm, while the zirconium atoms move outward along <111> by 0.018 nm. The oxygen atoms nearest to the zirconium, but not nearest to the vacancy, follow the displacement of the cation and move outwards along <111> by 0.004 nm. The oxygen atoms in the outermost right corner of the figure move inwards by 0.004 nm along <111>.

Tetragonal zirconia contains anion vacancies and may be written as ZrO_{2-x}, with x varying from 0.001 at 1,925°C to 0.052 at 2,410°C [6]. To accommodate these vacancies, surrounding ions move toward the vacancy to reduce its size. The two zirconium ions move by an approximate amount of 0.008 nm and the oxygen ions by 0.013 nm. The energy gain due to the relaxation of these ions is 0.22 eV. These values are, of course, different depending on the charge of the vacancy [25].

For the case of a singly-charged vacancy, the structural distortion results in the movement of surrounding oxygen ions by an approximate amount of 0.022 nm toward the vacancy and the zirconium ions by 0.009 nm away from the vacancy. These values are modified for the case of a doubly-charged vacancy to 0.033 nm for the oxygen ions and 0.022 nm for the zirconium ions. Obviously, the higher the positive charge of the vacancy, the greater the distortion towards or away from it. The energy gains due to the formation of singly- and doubly-charged vacancies are 1.0 and 3.3 eV, respectively.

Oxygen vacancies in monoclinic zirconia can occur in both the triple-planar and tetragonal geometries. When the vacancy is neutral, these vacancies have formation energies of 8.88 eV and 8.90 eV, respectively. Once the vacancy is singly charged positively (i.e., V^+) and in a tetrahedral position, the atomic relaxation energy is 0.47 eV. Creation of a doubly-charged positive vacancy (i.e., V^{2+}) in a tetrahedral position causes further displacement of the four surrounding zirconium ions away from the vacancy by an additional 0.01 nm. This leads to a further decrease in energy of 0.74 eV. Creation of a singly-charged negative vacancy (i.e., V^-) in the same tetrahedral position causes minimal displacement of the surrounding zirconium ions (by less than 0.002 nm) and an energy decrease that is less than 0.1 eV, which clearly points to the fact that the additional electron is only weakly localized in the vicinity of the vacancy and, hence, has little influence on the surrounding ions. The lattice relaxation and formation energies in the case of a neutral zirconium vacancy are about 1.4 and 24.2 eV, respectively. The oxygen ions surrounding this type of vacancy are displaced outwards from their equilibrium positions by about 0.01–0.02 nm.

At higher temperatures (i.e., 1,000°C) and excess partial pressure of oxygen (i.e., 10^{-6} to 1 atm.), monoclinic zirconia contains completely ionized zirconium vacancies [26]. At 1,000°C, zirconia is stoichiometric at a pressure of 10^{-16} atm. At this point, the concentration of oxygen vacancies is equal to twice the concentration of zirconium vacancies. As the partial pressure of oxygen increases, the stoichiometry changes such that for $ZrO_{2+\delta}$, with the δ value defined by:

$$\delta = 6 \times 10^{-3} \, p_{O_2}^{1/5},$$

(2)

where po_2 is the oxygen partial pressure in atm.

4 Mechanical Properties

Measurements of the mechanical properties of pure tetragonal and cubic zirconia are exceedingly difficult because of the higher temperatures required for such measurements. Hence, only monoclinic zirconia has been thoroughly studied in pure form. The mechanical properties of tetragonal and cubic zirconia have been determined for many stabilized zirconias and, because of the importance of these materials in engineering applications, several reviews have been written [27–29].

4.1 Elastic Properties

The measured elastic stiffness and compliance moduli for monoclinic zirconia have been summarized by Chan et al. [30]. The Young's and shear moduli of this same

Table 4 Polycrystalline Young's and shear moduli for monoclinic zirconia in GPa (adapted from Chan et al. [30]

	20°C	300°C	600°C	800°C	1,000°C
E_{Voigt}	266	256	250	245	239
E_{Reuss}	215	216	220	222	214
E_{Hill}	241	236	235	234	226
G_{Voigt}	104	99.1	96.8	94.9	92.6
G_{Reuss}	83.4	83.2	84.7	85.3	82.4
G_{Hill}	93.6	91.1	90.7	90.1	87.5

structure are given in Table 4 and were calculated using the Voigt, Reuss, and Hill approximations. The Voigt and Reuss approximations usually give the upper and lower bounds of these parameters. The maximum errors in these numbers are about 10% for most values, but can increase to greater than 20% for some of the transverse directions in the crystal.

For the monoclinic and tetragonal structures, the bulk modulus hovers around 150–200 GPa. Cubic zirconia has higher bulk modulus somewhere around 171–288 GPa. The high-pressure phases have values around 224–273 GPa and 254–444 GPa, for the orthorhombic-I and orthorhombic-II phases, respectively.

4.2 Hardness

The hardness for monoclinic zirconia is approximately 9.2 GPa [31] for samples with a density >98% and 4.1–5.2 GPa [32] for samples with a density >95% of theoretical, whereas hardness values for amorphous zirconia vary between 5 and 25 GPa [33]. The hardness increases slightly to values approaching 11 GPa for yttria-stabilized zirconia of 1.5 mol% yttria, which is stabilized in the tetragonal form [31]. Addition of larger amounts of yttria dopant results in hardness values approaching 15 GPa [34].

4.3 Toughness

The toughness of pure monoclinic zirconia is difficult to obtain because of problems encountered during sintering of these types of specimens. Generally, if a full density is desired for mechanical properties evaluation, the material needs to be heated to a temperature that is above the tetragonal-to-monoclinic transformation temperature (i.e., 1,471 K). This results in severe cracking upon cooling. However, there have been a few studies that have shown that nanocrystalline monoclinic zirconia can be sintered to full density at 1,273 K. In this case, microcracking during cooling can be avoided [35]. Unfortunately, these specimens have not been tested for toughness.

Experiments have been attempted with porous specimens of monoclinic zirconia and the fracture toughness has been extrapolated. A value of 2.06 ± 0.04 MPa m$^{1/2}$ was found for a specimen of 92.2 ± 0.4 % relative density, from which a fracture toughness of 2.6 MPa m$^{1/2}$ was extrapolated for a specimen of full density [36]. Slightly higher numbers of 3.7 ± 0.3 MPa•m$^{1/2}$ were found for specimens with >95% density [32]. Evidently, the fracture toughness of this phase of zirconia is quite low. The toughness of cubic zirconia is also low, reported as 2.8 MPa m$^{1/2}$ by Chiang et al. [37] and 1.8 ± 0.2 MPa•m$^{1/2}$ by Cutler et al. [32].

The addition of alloying elements such as Y^{3+}, Ce^{3+}, and Mg^{2+} can result in stabilization of tetragonal zirconia, which results in an increase in the fracture toughness of the material via a process of transformation toughening. The addition of increasing amounts of the stabilizing elements results in the stabilization of the cubic phase, which does not have transformation-toughening behavior. Toughening requires the presence of the metastable tetragonal phase.

As can be seen in Fig. 10 [29], the fracture toughness in polycrystalline tetragonal zirconia (TZP) and partially-stabilized zirconia (PSZ) appears to reach a maximum. This indicates a transition from flaw-size control of strength to transformation-limited strength. Ranges of fracture toughness values for zirconia composites are given by Richerson [38].

The stability of the tetragonal structure can be controlled by three factors: the grain size [39, 40], the constraint from a surrounding matrix [41, 42], and the amount of dopant additions. Commonly, very small tetragonal particles are added as a reinforcing phase to a matrix of another material, which is usually brittle (i.e., pure cubic or monoclinic zirconia, alumina [43], Si_3N_4 [44], and others [45]) as shown in Fig. 11a. This results in a higher overall toughness for the composite. For example, Gupta et al. [46] has shown that the addition of small tetragonal particles to a matrix of monoclinic zirconia results in an increment of the toughness to values between 6.07 and 9.07 MPa•m$^{1/2}$, in contrast to the low numbers observed for pure monoclinic zirconia. A review on the transformation toughnening of several zirconia composites has been prepared by Bocanegra-Bernal and Diaz De La Torre [42].

This toughening mechanism is associated with the increase in volume upon transformation to the monoclinic phase. Since the monoclinic phase occupies a larger volume compared with the tetragonal phase, it forces closure of any propagating cracks, greatly diminishing the catastrophic failure of the material due to fracture [47]. In addition, the transformation from tetragonal to monoclinic results in energy absorption that blunts the crack.

The transformation is induced by an applied stress on the material. Initially, a ceramic composite may contain a crack that begins to propagate upon application of

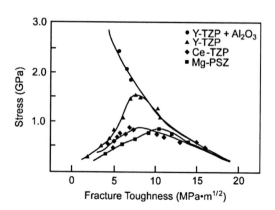

Fig. 10 Strength vs. fracture toughness for a selection of ZrO_2-toughened engineering ceramics [29] (reprinted with permission)

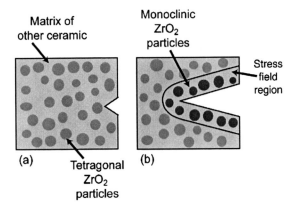

Fig. 11 Transformation toughening mechanism in a composite that contains small tetragonal zirconia particles

a stress (Fig. 11a). If the composite contains metastable tetragonal particles, the large stresses at the tip of the crack can force the tetragonal-to-monoclinic transformation of these particles increasing the volume of material in the region of the crack and forcing crack closure (Fig. 11b). The positive change in volume during the transformation is small but significant. If the positive change in volume is large, it can result in fragmentation of the material. On the other hand, a negative change in volume will not result in strains that promote crack closure. Hence, zirconia is quite unique in that the monoclinic and tetragonal structures are very close in density such that exaggerated volume increases are avoided during transformation. As the amount of dopant is increased, the stability of the tetragonal phase is higher and the transformation becomes more sluggish.

Indeed, Bravo-Leon et al. [31] and Sakuma et al. [34] have found that the toughness is higher for samples with yttria concentrations lower than the typical 3 mol% used for this material. Fracture toughness values of 16–17 MPa m$^{1/2}$ were reached by Bravo-Leon et al. for a 1 mol% yttria specimen with a grain size of 90 nm and a 1.5 mol% yttria specimen with a grain size of 110 nm. This can be attributed to the lower stability of the tetragonal phase with low dopant concentrations, which easily transforms to the monoclinic phase upon application of the stress.

4.4 Creep

Using the strain rate data shown in Fig. 12, the activation energy for creep in monoclinic zirconia has been found to be $Q_c \approx 330$–360 kJ mol^{-1} [48, 49]. The measured stress exponent, n, from equation:

$$\dot{\varepsilon} = A \frac{Gb}{kT} \left(\frac{b}{d} \right)^p \left(\frac{\sigma}{G} \right)^n D, \tag{3}$$

was found to be 1.7 by Roddy et al. [48] and 2.3–2.5 by Yoshida et al. [49]. In this equation, A is a constant, G is the shear modulus, b is the Burger's vector, d is the grain size,

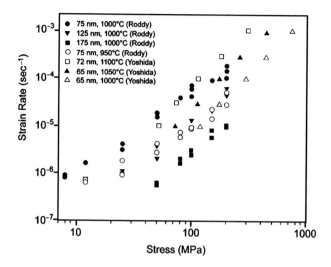

Fig. 12 Stress dependence of monoclinic zirconia during creep deformation (adapted from Roddy et al. [48] and Yoshida et al. [49])

σ is the applied stress, p is the grain size exponent, n is the stress exponent, and D is the diffusion coefficient. The value by Roddy et al. is intermediate between diffusional creep ($n = 1$) and superplastic deformation ($n = 2$), but closer to superplastic deformation, whereas the values by Yoshida et al. are higher than both, but close to the stress exponent for superplastic deformation. The grain size exponent, p, was found to be 2.8 by Roddy et al., which is closer to $p = 3$ for Coble creep (lattice diffusion) than $p = 2$ for superplastic or Nabarro-Herring creep (grain boundary diffusion). However, Yoshida et al. found values between 2.4 and 2.5. From the exponents found in both studies, it is likely that creep deformation in monoclinic zirconia is due to superplastic deformation.

5 Electronic Properties

Cubic zirconia doped with oxides such as Y_2O_3 or CaO is the material of choice for many high temperature applications because of its extremely high ionic conductivity at intermediate and high temperatures. A review on the properties of these specialized rare-earth stabilized zirconia materials has been prepared by Comins et al. [50].

The oxygen pressure dependence of the conductivity in tetragonal zirconia can be seen in Fig. 13 [51]. This material is a mixed electronic and ionic conductor with a large ionic contribution except at very high temperatures or very low oxygen partial pressures. The electronic component of the conductivity arises from doubly-charged oxygen vacancies at lower oxygen pressures and a temperature of 1,400°C. Other contributions to conductivity are difficult to determine. The movement of oxygen vacancies can take place along two directions for the tetragonal structure: within the x–y plane along the [110] direction or perpendicular to this plane along the [001] direction. In both directions, the O–O distances are very similar (0.2640 nm within the (x, y) plane and 0.2644 nm in the direction perpendicular to that plane) [25]. From these numbers, it would appear that there is no preferential direction for diffusion.

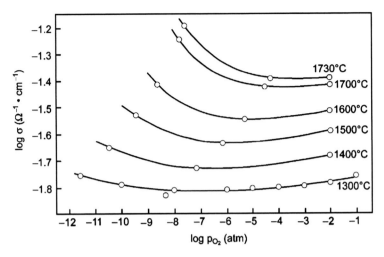

Fig. 13 Conductivity isotherms for tetragonal zirconia as a function of oxygen pressure [51] (reprinted with permission)

However, the diffusion process is controlled by the Zr–Zr distance and not by the O–O distance, since the vacancy must move between two such ions to diffuse. Along the two relevant directions, these distances are 0.3655 nm for the [110] direction and 0.3645 nm for the [001] direction. The diffusion barriers for movement of a neutral vacancy along [110] and [001] are 1.35 and 1.43 eV, respectively. This is expected from the fact that there is a smaller gap between zirconium ions along the [001] direction. Hence, diffusion along this direction proves to be more difficult. The diffusion barriers for movement of a doubly-charged vacancy along the two relevant directions are 0.22 and 0.61 eV, respectively. Again, movement along the [001] direction proves to be more difficult. This can be visualized in Fig. 14 [6].

Monoclinic zirconia is both an electron and ion conductor depending on the temperature and oxygen pressure (Fig. 15) [52–54]. At low pressures, it exhibits n-type behavior in which the charge carriers are double-charged oxygen vacancies, while at higher pressures it exhibits p-type behavior in which the charge carriers are singly-ionized oxygen interstitials. The transition from n-type to p-type is established by the change in sign of the conductivity curve. Assuming the -1/6 and 1/5 dependences in the two regions are good fits to the data, the total conductivity at 1,000°C can be represented by:

$$\sigma_{1,000°C} \cong 8.5 \times 10^{-5} \, p_{O_2}^{1/5} + 1.1 \times 10^{-9} \, p_{O_2}^{-1/6} + 3.2 \times 10^{-6}. \tag{4}$$

In addition, Vest et al. [53] determined the hole mobility at 1,000°C to be $\mu_{1,000°C} = 1.4 \times 10^{-6} \, cm^2 \cdot V^{-1} \cdot s^{-1}$.

If the pressure is kept constant and the temperature is increased, the conductivity also increases (see Fig. 4 of Kumar et al. [52]). At lower temperatures (<600°C), conductivity is predominantly ionic, and at higher temperatures (>700°C), it is predominately electronic. Between 600 and 700°C, both ionic and electronic conductivities are seen in this material. Values of the activation energies required for each type of

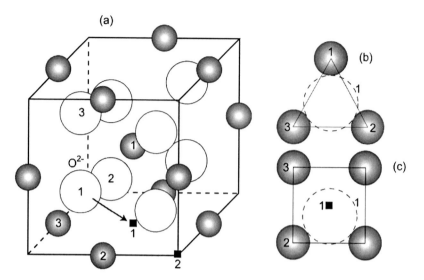

Fig. 14 Simplified representation of possible diffusion mechanism for oxygen atoms in tetragonal zirconia. (**a**) Tetragonal cell with two octahedral empty sites, marked with *black squares* 1 and 2, (**b**) Position of oxygen 1 during its motion past the zirconium 1–3 face, (**c**) possible off-centered position for oxygen 1 inside octahedral site 1 (adapted from [6])

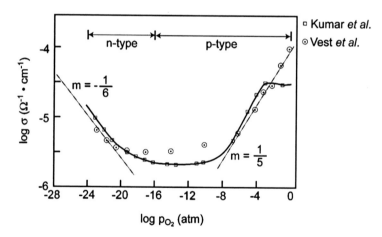

Fig. 15 Oxygen pressure dependence of total conductivity for monoclinic zirconia at 990°C (adapted from Kumar et al. [52] and Vest et al. [53])

conductivity are still a matter of controversy because of the complexity of the conduction processes. Earlier values include numbers such as 3.56 eV for n-type conductivity and 0.86 eV for p-type conductivity [55].

The conductivity of two high-pressure phases of zirconia is shown in Fig. 16 [56]. The discontinuities in the conductivity occur approximately at 1,000°C for the sample at 16.5 GPa and 1,050°C for the sample at 18.0 GPa. At the higher temperatures, the conductivity corresponds to a so-called "cubic" high-pressure and high-temperature phase of zirconia, although its exact nature was not determined by Ohtaka et al. [56].

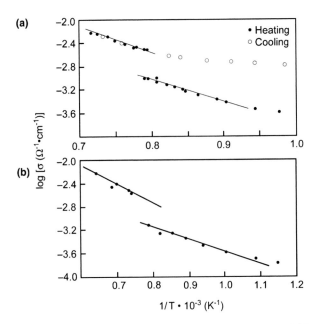

Fig. 16 Electrical conductivity of pure cubic zirconia at (a) 16.5 GPa and (b) 18 GPa [56] (reprinted with permission)

At the lower temperatures, the conductivity corresponds to the orthorhombic-II phase. From the Arrhenius plots in the figure, approximate activation energies for conduction can be obtained. For the "cubic" phase, the activation energies are 8.80 and 0.60 eV at pressures of 16.5 and 18.0 GPa, respectively, while for the orthorhombic-II phase they are 0.72 and 0.40 eV for the two pressures studied.

6 Diffusion Coefficients

Diffusion in zirconia is closely linked to ionic conductivity. Consequently, some diffusion data has already been presented in Sect. 5. This section will include additional results particularly for monoclinic zirconia. Oxygen self-diffusion at a pressure of 300 Torr, as determined by testing zirconia spheres of diameters between 75 and 105 μm, behaves as shown in Fig. 17 [57], where D is the diffusion coefficient, t is time, and a is the sphere radius. At a pressure of 700 Torr, the behavior changes to that shown in Fig. 18 [58]. In this case D^* is the self-diffusion coefficient and the rest of the terms are as defined before, with $a = 100–150$ μm. Both of these experiments were performed in an oxygen atmosphere of $^{18}O–^{16}O$. The self-diffusion coefficients calculated from the diffusion data obey Arrhenius expressions as illustrated in Fig. 19 [57, 58]. The linear fits describing the diffusion coefficient at 300 and 700 Torr, are given by:

$$P = 300 \text{ Torr} : D\left(\frac{cm^2}{s}\right) = 9.73 \pm 1.4 \times 10^{-3} \exp\left\{-\frac{56.0 \pm 2.4 \text{ kcal/mol}}{RT}\right\} \quad (5)$$

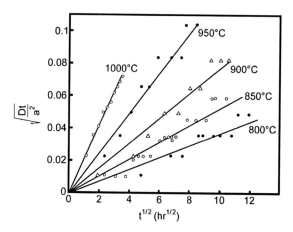

Fig. 17 Oxygen diffusion in monoclinic zirconia spheres at 300 Torr [57] (reprinted with permission)

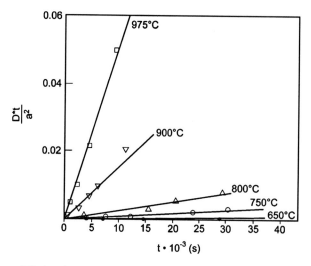

Fig. 18 Oxygen diffusion in monoclinic zirconia spheres at 700 Torr [58] (reprinted with permission)

$$P = 700 \text{ Torr}: D\left(\frac{\text{cm}^2}{s}\right) = 2.34 \times 10^{-2} \exp\left\{-\frac{45300 \pm 1200 \text{ cal/mol}}{RT}\right\}. \tag{6}$$

According to Ikuma et al. [59], surface diffusion and lattice diffusion should be separated and result in the following diffusion coefficients:

$$D_{surface}\left(\text{cm}^2/s\right) = 5.84 \times 10^{-12} \exp\left\{-\frac{75.3 \text{ kJ/mol}}{RT}\right\} \tag{7}$$

Fig. 19 Arrhenius plot of oxygen self-diffusion in monoclinic zirconia (adapted from Madeyski and Smeltzer
[57] and Keneshea and Douglass [58])

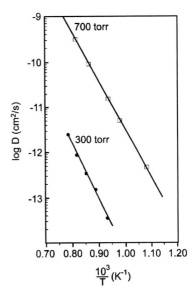

$$D_{lattice}\left(cm^2/s\right) = 4.84 \times 10^{-12} \exp\left\{-\frac{87.0 \text{ kJ/mol}}{RT}\right\}. \tag{8}$$

These two expressions are not that very different. Hence, the macroscopic diffusion
behavior of monoclinic zirconia can be approximated by lattice diffusion, while surface diffusion can be ignored.

Diffusion in pure tetragonal and cubic zirconia is experimentally challenging
because it requires the higher temperatures at which the two phases are stable.
However, simulations at temperatures between 1,273 and 2,673 K have been performed
on cubic zirconia, showing noticeable, but not large, oxygen ion diffusion along the
grain boundaries and a significant energy barrier to movement from the grain boundaries into the bulk, although at higher temperatures diffusion is obviously enhanced.
However, even at higher temperatures, diffusion along the grain boundary is not as
favorable as that across the grain boundary [60].

7   Phase Transitions and the Processing of Zirconia

Upon heating, the monoclinic phase in zirconia starts transforming to the tetragonal
phase at 1,461 K, peaks at 1,471 K, and finishes at 1,480 K. On cooling, the transformation from the tetragonal to the monoclinic phase starts at 1,326 K, peaks at 1,322 K,
and finishes at 1,294 K, exhibiting a hysteresis behavior that is well known for this
material [61–65]. This transformation can also be affected by irradiation with heavy
ions, such as 300 MeV Ge [66] and 340 keV Xe [67].

The tetragonal phase transforms to the cubic fluorite structure at 2584 ± 15 K [68].
This transformation temperature has been found to be dependent on the atmosphere

in which the transformation is taking place [6]. In a reducing atmosphere, the transformation takes place at approximately 2,323 K, and in a neutral atmosphere, it takes place at approximately 2,563 K, which is in proximity to the highly accurate value found by Navrotsky et al. [68]. Continued heating of the material results in melting at a temperature of 2,963 K [63]. The phase stability as a function of pressure for this material in its pure form is shown in Fig. 20 [7].

The practical use of pure zirconia is restricted by the monoclinic to tetragonal transformation, as this transformation causes cracking and sometimes complete disintegration of the specimen. Depending on the orientation of the particular grain that is undergoing the transformation, there is a maximum strain in the lattice of ~4% [29], which is quite significant and promotes failure of the specimen when undergoing heating and cooling cycles.

This transformation has many characteristics of martensitic transformations in metals, with definite orientation relationships between the two structures. The orientation relationships conform to the following [69–71]:

$$
\left.
\begin{aligned}
&(100)_m \,\|(1\overline{1}0)_{bct} \text{ and } [010]_m \,\|[001]_{bct}, \\
&\text{and by twinning } (100)_m \,\|(110)_{bct} \text{ and } [001]_m \,\|[001]_{bct}
\end{aligned}
\right\}
\tag{9}
$$

where m and t represent the monoclinic and tetragonal phases, and bct refers to the body-centered tetragonal structure. Possible variants of these twin relationships for small tetragonal particles are shown in Fig. 21. In this figure, the hashed areas represent the transformed monoclinic phase and the unhashed areas represent the

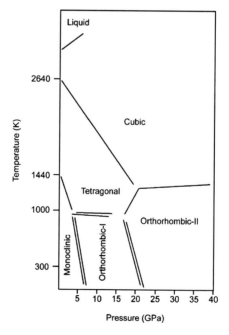

Fig. 20 Pressure–temperature phase diagram of zirconia [7] (reprinted with permission)

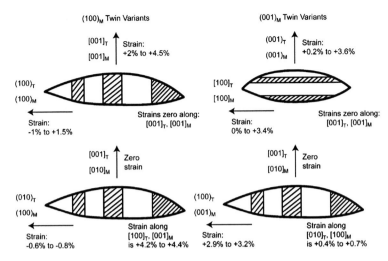

Fig. 21 The four possible arrangements of twin-related variants together with the range of strain values predicted for the directions indicated (adapted from Kelly [76])

untransformed tetragonal phase. As the transformation progresses, the entire particle eventually forms the stable monoclinic phase for this material. The transformation progresses in two stages. The first stage involves a displacive transformation with small shifts of the atoms and the second stage involves a martensitic transformation in which both structures remain almost unchanged [72]. It is this latter transformation that has been studied the most thoroughly [73–75].

To avoid this destructive transformation, stabilization of the tetragonal and cubic structures of zirconia can be done at room temperature by the addition of trivalent dopant ions such as Y^{3+} and Ce^{3+}, divalent dopant ions such as Ca^{2+}, or tetravalent dopant ions. Doping of zirconia has enormous consequences not only for the mechanical properties of this material, but also for the electronic properties. In particular, Y^{3+} has a large solubility range in zirconia and can be used to stabilize both the tetragonal and cubic phases. To maintain charge neutrality, one oxygen vacancy must be created for each pair of dopant cations that are added to the structure. This results in large increases in ionic conductivity. Stabilization of the tetragonal and cubic structures requires differing amounts of dopants. The tetragonal phase is stabilized at lower dopant concentrations. The cubic phase is stabilized at higher dopant concentrations, as shown in the room temperature region of the ZrO_2–Y_2O_3 phase diagram in Fig. 22 [77]. At higher Y_2O_3 doping, the material exhibits an ordered $Zr_3Y_4O_{12}$ phase at 40 mol% Y_2O_3, a eutectoid at a temperature < 400°C at a composition between 20 and 30 mol% Y_2O_3, a eutectic at 83 ± 1 mol% Y_2O_3, and a peritectic at 76 ± 1 mol% Y_2O_3 [78]. Other zirconia phase diagrams have been developed by Stubican and Ray for ZrO_2–CaO [79], Grain for ZrO_2–MgO [80], Cohen and Schaner for ZrO_2–UO_2 [81], Mumpton and Roy for ZrO_2–ThO_2 [82], Barker et al [83] for ZrO_2–Sc_2O_3, and Duwez and Odell for ZrO_2–CeO_2 [84], among others.

As mentioned briefly in Sect. 4, another way of stabilizing the tetragonal structure at room temperature is the formation of nanocrystalline powders or nanograined sintered specimens. To obtain powders of dense PSZ compacts at room temperature, the material has to contain crystals or grains below a certain critical size, which

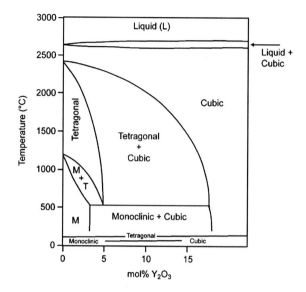

Fig. 22 Zirconia-rich end of the yttria-zirconia phase equilibrium diagram [77] (reprinted with permission)

increases as the dopant concentration increases. The critical size is 22.6 (also found to be ~18 nm by Chraska et al. [85] and 15.3 nm by Garvie [86]), 41.7, 67, and 93.8 nm for yttria doping concentrations of 0, 0.5, 1.0, and 1.5 mol% [87]. The values decrease with increasing dopant concentration, consistent with the fact that yttria is a tetragonal-phase stabilizer. Changes in the transformation temperature with dopant concentration and crystallite size are shown in Fig. 23 [87], where it can be seen that the transformation temperature decreases with decreasing crystallite size and increasing dopant concentration. The dotted lines represent theoretical curves calculated according to:

$$
T_{\text{transformation}} = \frac{\Delta H_{\text{vol}} + \dfrac{10\Delta h_{\text{surf}}}{d_{\text{critical}}}}{\Delta S_{\text{vol}} + \dfrac{10\Delta s_{\text{surf}}}{d_{\text{critical}}}},
\tag{10}
$$

where ΔH_{vol} is the volumetric heat of transformation, Δh_{surf} is the surface enthalpy difference, d_{critical} is the critical crystallite size to stabilize the tetragonal phase at room temperature, ΔS_{vol} is the volumetric entropy of transformation, and Δs_{surf} is the surface entropy difference. The solid curves are from the standard ZrO_2–Y_2O_3 phase diagram (Fig. 22). The solid circles represent experimental data on samples that happened to have crystallite sizes close to those for which the theoretical curves were calculated.

The stabilization of the tetragonal phase at room temperature due to a decrease in the crystallite size has been attributed to a surface energy difference and roughly obeys the relationships [88]:

$$
\frac{1}{d_{\text{critical}}} = -\frac{\Delta H_\infty T}{6\Delta\gamma T_b} + \frac{\Delta H_\infty}{6\Delta\gamma} \text{ (for powders)}
\tag{11}
$$

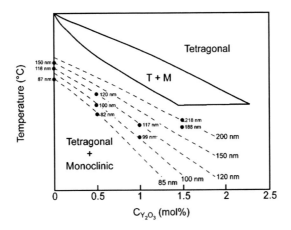

Fig. 23 Phase diagram representation of the crystallite size and yttria concentration dependency of the tetragonal-to-monoclinic transformation temperature [87] (reprinted with permission)

$$\frac{1}{d_{critical}} = -\frac{\Delta H_\infty T}{6\Delta\Sigma T_b} + \frac{\Delta H_\infty + \Delta U_{sc}}{6\Delta\Sigma} \text{ (for sintered pellets)} \qquad (12)$$

where $d_{critical}$ is the critical crystallite/grain size, ΔH_∞ is the enthalpy of the tetragonal-to-monoclinic phase transformation in a sample with infinite crystallite/grain size, T is the temperature of transformation, $\Delta\gamma$ is the difference in surface energy in powder crystallites, $\Delta\Sigma$ is the difference in interfacial energy in sintered pellets, T_b is the transformation temperature for an infinitely large-grained sample, and ΔU_{sc} is the strain energy involved in the transformation. From these equations, it can be seen that the same material in the solid form has a lower transformation temperature than in the powder form. This difference is due to the strain energy, ΔU_{sc}, involved in the transformation, which is present only in the pellets since there is a requirement for geometric compatibility that is not present in the powders.

References

1. G. Stapper, M. Bernasconi, N. Nicoloso, and M. Parrinello, Ab initio study of structural and electronic properties of yttria-stabilized cubic zirconia, *Phys. Rev. B*, **59**(2), 797–810 (1999).
2. J.K. Dewhurst and J.E. Lowther, Relative stability, structure, and elastic properties of several phases of pure zirconia, *Phys. Rev. B*, **57**(2), 741–747 (1998).
3. P. Li, I.-W. Chen, and J.E. Penner-Hahn, Effect of dopants on zirconia stabilization – An X-ray absorption study: I, trivalent dopants, *J. Am. Ceram. Soc.* **77**(1) 118–128 (1994).
4. D.W. Richerson, *Modern Ceramic Engineering Properties, Processing, and Use in Design 3/e*, Taylor and Francis Group, Boca Raton, 2006, p 30.
5. S.-M. Ho, On the structural chemistry of zirconium oxide, *Mater. Sci. Eng.* **54**, 23–29 (1982).

6. P. Aldebert and J.-P. Traverse, Structure and ionic mobility of zirconia at high temperature, *J. Am. Ceram. Soc.* **68**(1), 34–40 (1985).
7. P. Bouvier, E. Djurado, G. Lucazeau, and T. Le Bihan, High-pressure structural evolution of undoped tetragonal nanocrystalline zirconia, *Phys. Rev. B* **62**(13), 8731–8737 (2000).
8. N. Igawa and Y. Ishii, Crystal structure of metastable tetragonal zirconia up to 1473 K, *J. Am. Ceram. Soc.* **84**(5), 1169–1171 (2001).
9. G. Teufer, The crystal structure of tetragonal ZrO_2, *Acta Crystallogr.* **15**, 1187 (1962).
10. M. Winterer, Reverse Monte Carlo analysis of extended x-ray absorption fine structure spectra of monoclinic and amorphous zirconia, *J. Appl. Phys.* **88**(10), 5635–5644 (2000).
11. P. Li, I.-W. Chen, and J.E. Penner-Hahn, X-ray-absorption studies of zirconia polymorphs. I. Characteristic local structures, *Phys. Rev. B*, **48**(14), 10063–10073 (1993).
12. J.D. McCullough and K.N. Trueblood, The crystal structure of baddeleyite (monoclinic ZrO2), *Acta Crystallogr.* **12**, 507–511 (1959)
13. H. Arashi, T. Suzuki, ad S.-I. Akimoto, Non-destructive phase transformation of ZrO_2 single crystal at high pressure, *J. Mater. Sci. Lett.* **6**, 106–108 (1987).
14. S. Kawasaki, T. Yamanaka, S. Kume, and T. Ashida, Crystallite size effect on the pressure-induced phase transformation of ZrO_2, *Solid State Commun.* **76**(4), 527–530 (1990).
15. S. Desgreniers and K. Lagarec, High-density ZrO_2 and HfO_2: Crystalline structures and equations of state, *Phys. Rev. B*, **59**(13), 8467–8472 (1999).
16. C.J. Howard, E.H. Kisi, and O. Ohtaka, Crystal structures of two orthorhombic zirconias, *J. Am. Ceram. Soc.* **74**(9), 2321–2323 (1991).
17. E.H. Kisi, C.J. Howard, and R.J. Hill, Crystal structure of orthorhombic zirconia in partially stabilized zirconia, *J. Am. Ceram. Soc.* **72**(9), 1757–1760 (1989).
18. O. Ohtaka, T. Yamanaka, and T. Yagi, New high-pressure and –temperature phase of ZrO_2 above 1000°C at 20 GPa, *Phys. Rev. B*, **49**(14), 9295–9298 (1994).
19. J. Livage, K. Doi, and C. Mazieres, Hydrated zirconium oxide, *J. Am. Ceram. Soc.* **51**(6), 349–353 (1968).
20. A. Clearfield, Crystalline hydrous zirconia, *Inorg. Chem.* **3**(1), 146–148 (1964).
21. C. Landron, A. Douy, and D. Bazin, From liquid to solid: residual disorder in the local environment of oxygen-coordinated zirconium, *Phys. Status Solidi B*, **184**, 299–307 (1994).
22. A. Corina Geiculescu and H.J. Rack, Atomic-scale structure of water-based zirconia xerogels by X-ray diffraction, *J. Sol–Gel Sci. Technol.* **20**, 13–26 (2001).
23. A.V. Chadwick, G. Mountjoy, V.M. Nield, I.J.F. Poplett, M.E. Smith, J.H. Strange, and M.G. Tucker, Solid-state NMR and X-ray studies of the structural evolution of nanocrystalline zirconia, *Chem. Mater.* **13**, 1219–1229 (2001).
24. A.S. Foster, V.B. Sulimov, F. Lopez Gejo, A.L. Shluger, and R.M. Nieminen, Structure and electrical levels of point defects in monoclinic zirconia, *Phys. Rev. B*, **64**, 224108 (2001).
25. A. Eichler, Tetragonal Y-doped zirconia: structure and ion conductivity, *Phys. Rev. B*, **64**, 174103 (2001).
26. R.W. Vest, N.M. Tallan, and W.C. Tripp, Electrical properties and defect structure of zirconia: I, monoclinic phase, *J. Am. Ceram. Soc.* **47**(12), 635–640 (1964).
27. A.H. Heuer, Transformation toughening in ZrO_2-containing ceramics, *J. Am. Ceram. Soc.* **70**(10), 689–698 (1987).
28. U. Messerschmidt, B. Baufeld, and D. Baither, Plastic deformation of cubic zirconia single crystals, *Key Eng. Mater.* **153–154**, 143–182 (1998).
29. R.H.J. Hannink, P.M. Kelly, and B.C. Muddle, Transformation toughening in zirconia-containing ceramics, *J. Am. Ceram. Soc.* **83**(3), 461–487 (2000).
30. S.-K. Chan, Y. Fang, M. Grimsditch, Z. Li, M.V. Nevitt, W.M. Robertson, and E.S. Zouboulis, Temperature dependence of the elastic moduli of monoclinic zirconia, *J. Am. Ceram. Soc.* **74**(7), 1742–1744 (1991).
31. A. Bravo-Leon, Y. Morikawa, M. Kawahara, and M.J. Mayo, Fracture toughness of nanocrystalline tetragonal zirconia with low yttria content, *Acta Mater.* **50**, 4555–4562 (2002).
32. R.A. Cutler, J.R. Reynolds, and A. Jones, Sintering and characterization of polycrystalline monoclinic, tetragonal, and cubic zirconia, *J. Am. Ceram. Soc.* **75**(8), 2173–2183 (1992).

33. M. Levichkova, V. Mankov, N. Starbov, D. Karashanova, B. Mednikarov, and K. Starbova, Structure and properties of nanosized electron beam deposited zirconia thin films, *Surf. Coat. Technol.* **141**, 70–77 (2001).

34. T. Sakuma, Y.-I. Yoshizawa, and H. Suto, The microstructure and mechanical properties of yttria-stabilized zirconia prepared by arc-melting, *J. Mater. Sci.* **20**, 2399–2407 (1985).

35. G. Skandan, H. Hahn, M. Roddy, and W.R. Cannon, Ultrafine-grained dense monoclinic and tetragonal zirconia, *J. Am. Ceram. Soc.* **77**(7), 1706–1710 (1994).

36. J. Eichler, U. Eisele, and J. Rödel, Mechanical properties of monoclinic zirconia, *J. Am. Ceram. Soc.* **87**(7), 1401–1403 (2004).

37. Y.-M. Chiang, D. Birnie III, and W.D. Kingery, *Physical Ceramics Principles for Ceramic Science and Engineering*, Wiley, New York, 1997, p. 484.

38. D.W. Richerson, *Modern Ceramic Engineering Properties, Processing, and Use in Design 3/e* Taylor and Francis Group, Boca Raton, 2006, pp. 275, 643.

39. G. Stefanic and S. Music, Factors influencing the stability of low temperature tetragonal ZrO_2, *Croat. Chem. Acta* **75**(3), 727–767 (2002).

40. H.S. Maiti, K.V.G.K. Gokhale, and E.C. Subbarao, Kinetics and burst phenomenon in ZrO_2 transformation, *J. Am. Ceram. Soc.* **55**(6), 317–322 (1972).

41. A.H. Heuer, N. Claussen, W.M. Kriven, and M. Rühle, Stability of tetragonal ZrO_2 particles in ceramic matrices, *J. Am. Ceram. Soc.* **65**(12), 642–650 (1982).

42. M.H. Bocanegra-Bernal and S. Diaz De La Torre, Review. Phase transitions in zirconium dioxide and related materials for high performance engineering materials, *J. Mater. Sci.* **37**, 4947–4971 (2002).

43. A.G. Evans, N. Burlingame, M. Drory, and W.M. Kriven, Martensitic transformations in zirconia–particle size effects and toughening, *Acta Metall.* **29**, 447–456 (1981).

44. A.G. Evans and A.H. Heuer, Review – Transformation toughening in ceramics: Martensitic transformations in crack-tip stress fields, *J. Am. Ceram. Soc.* **63**(5–6), 241–248 (1980).

45. D.W. Richerson, *Modern Ceramic Engineering Properties, Processing, and Use in Design* Taylor and Francis Group, Boca Raton, 2006, pp. 635, 640–644.

46. T.K. Gupta, F.F. Lange, and J.H. Bechtold, Effect of stress-induced phase transformation on the properties of polycrystalline zirconia containing metastable tetragonal phase, *J. Mater. Sci.* **13**(7), 1464–1470 (1978).

47. Y.-M. Chiang, D. Birnie III, and W.D. Kingery, *Physical Ceramics Principles for Ceramic Science and Engineering*, Wiley, New York, 1997, pp. 488–492.

48. M.J. Roddy, W.R. Cannon, G. Skandan, and H. Hahn, Creep behavior of nanocrystalline monoclinic ZrO_2, *J. Eur. Ceram. Soc.* **22**, 2657–2662 (2002).

49. M. Yoshida, Y. Shinoda, T. Akatsu, and F. Wakai, Superplasticity-like deformation of nanocrystalline monoclinic zirconia at elevated temperatures, *J. Am. Ceram. Soc.* **87**(6), 1122–1125 (2004).

50. J.D. Comins, P.E. Ngoepe, and C.R.A. Catlow, Brillouin-scattering and computer-simulation studies of fast-ion conductors. A review, *J. Chem. Soc. Faraday Trans.* **86**(8), 1183–1192 (1990).

51. R.W. Vest and N.M. Tallan, Electrical properties and defect structure of zirconia: II, tetragonal phase and inversion, *J. Am. Ceram. Soc.* **48**(9), 472–475 (1965).

52. A. Kumar, D. Rajdev, and D.L. Douglass, Effect of oxide defect structure on the electrical properties of ZrO_2, *J. Am. Ceram. Soc.* **55**(9), 439–445 (1972).

53. R.W. Vest, N.M. Tallan, and W.C. Tripp, Electrical properties and defect structure of zirconia: I, monoclinic phase, *J. Am. Ceram. Soc.* **47**(12), 635–640 (1964).

54. P. Kofstad and D.J. Ruzicka, On the defect structure of ZrO_2 and HfO_2, *J. Electrochem. Soc.* **110**(3), 181–184 (1963).

55. E. Dow Whitney, Electrical resistivity and diffusionless phase transformation of zirconia at high temperatures and ultrahigh pressures, *J. Electrochem. Soc.* **112**(1), 91–94 (1965).

56. O. Ohtaka, S. Kume, and E. Ito, Stability field of cotunnite-type zirconia, *J. Am. Ceram. Soc.* **73**(3), 744–745 (1990).

57. A. Madeyski and W.W. Smeltzer, Oxygen diffusion in monoclinic zirconia, *Mater. Res. Bull.* **3**, 369–376 (1968).

58. F.J. Keneshea and D.L. Douglas, The diffusion of oxygen in zirconia as a function of oxygen pressure, *Oxidation Met.* **3**(1), 1–14 (1971).
59. Y. Ikuma, K. Komatsu, and W. Komatsu, Oxygen diffusion in monoclinic ZrO_2, undoped and doped with Y_2O_3, *Adv. Ceram.* **24**, 749–758 (1988).
60. C.A.J. Fisher and H. Matsubara, Molecular dynamics investigations of grain boundary phenomena in cubic zirconia, *Comput. Mater. Sci.* **14**, 177–184 (1999).
61. Y. Moriya and A. Navrotsky, High-temperature calorimetry of zirconia: heat capacity and thermodynamics of the monoclinic-tetragonal phase transition, *J. Chem. Thermodyn.* **38**, 211–223 (2006).
62. H. Boysen, F. Frey, and T. Vogt, Neutron powder investigation of the tetragonal to monoclinic phase transformation in undoped zirconia, *Acta Crystallogr. B*, **47**, 881–886 (1991).
63. R. Ruh, H.J. Garrett, R.F. Domagala, and N.M. Tallan, The system zirconia-hafnia, *J. Am. Ceram. Soc.* **51**(1), 23–27 (1968).
64. G.M. Wolten, Diffusionless phase transformations in zirconia and hafnia, *J. Am. Ceram. Soc.* **46**(9), 418–422 (1963).
65. W.L. Baun, Phase transformation at high temperatures in hafnia and zirconia, *Science*, **140**(3573), 1330–1331 (1963).
66. A. Benyagoub, F. Levesque, F. Couvreur, C. Gibert-Mougel, C. Dufour, and E. Paumier, Evidence of a phase transition induced in zirconia by high energy heavy ions, *Appl. Phys. Lett.* **77**(20), 3197–3199 (2000).
67. K.E. Sickafus, H. Matzke, T. Hartmann, K. Yasuda, J.A. Valdez, P. Chodak III, M. Nastasi, and R.A. Verrall, Radiation damage effects in zirconia, *J. Nucl. Mater.* **274**, 66–77 (1999).
68. A. Navrotsky, L. Benoist, and H. Lefebvre, Direct calorimetric measurement of enthalpies of phase transitions at 2000–2400°C in yttria and zirconia, *J. Am. Ceram. Soc.* **88**(10), 2942–2944 (2005).
69. J.E. Bailey, The monoclinic-tetragonal transformation and associated twinning in thin films of zirconia, *Proc. Roy. Soc. Lon. Ser. A, Math. Phys. Sci.* **279**(1378), 395–412 (1964).
70. S.T. Buljan, H.A. McKinstry, and V.S. Stubican, Optical and X-ray single crystal studies of the monoclinic tetragonal transition in ZrO_2, *J. Am. Ceram. Soc.* **59**(7–8), 351–354 (1976).
71. R.N. Patil and E.C. Subbarao, Monoclinic-tetragonal phase transition in zirconia: Mechanism, pretransformation and coexistence, *Acta Crystallogr. A* **26**, 535–542 (1970).
72. F. Frey, H. Boysen, and T. Vogt, Neutron powder investigation of the monoclinic to tetragonal phase transformation in undoped zirconia, *Acta Crystallogr. B* **46**, 724–730 (1990).
73. E.C. Subbarao, H.S. Maiti, and K.K. Srivastava, Martensitic transformation in zirconia, *Phys. Status Solidi A* **21**, 9–40 (1974).
74. G.K. Bansal and A.H. Heuer, On a martensitic phase transformation in zirconia (ZrO_2)-I. Metallographic evidence, *Acta Metall.* **20**, 1281–1289 (1972).
75. G.K. Bansal and A.H. Heuer, On a martensitic phase transformation in zirconia (ZrO_2)-II. Crystallographic aspects, *Acta Metall.* **22**, 409–417 (1974).
76. P.M. Kelly, Martensitic transformations in ceramics, *Mater. Sci. Forum*, **56–58**, 335–346 (1990).
77. D. Huang, K.R. Venkatachari, and G.C. Stangle, Influence of yttria content on the preparation of nanocrystalline yttria-doped zirconia, *J. Mater. Res.* **10**(3), 762–773 (1995).
78. V.S. Stubican, R.C. Hink, and S.P. Ray, Phase equilibria and ordering in the system ZrO_2-Y_2O_3, *J. Am. Ceram. Soc.* **61**(1–2), 17–21 (1978).
79. V.S. Stubican and S.P. Ray, Phase equilibria and ordering in the system ZrO_2-CaO, *J. Am. Ceram. Soc.* **60**(11–12), 534–537 (1977).
80. C.F. Grain, Phase relations in the ZrO_2-MgO system, *J. Am. Ceram. Soc.* **50**(6), 288–290 (1967).
81. I. Cohen and B.E. Schaner, A metallographic and x-ray study of the UO_2-ZrO_2 system, *J. Nucl. Mater.* **9**(1), 18–52 (1963).
82. F.A. Mumpton and R. Roy, Low-temperature equilibria among ZrO_2, ThO_2, and UO_2, *J. Am. Ceram. Soc.* **43**, 234–240 (1960).
83. W.W. Barker, F.P. Bailey, and W. Garrett, A high-temperature neutron diffraction study of pure and scandia-stabilized zirconia, *J. Solid State Chem.* **7**, 448–453 (1973).
84. P. Duwez and F. Odell, Phase relationships in the system zirconia-ceria, *J. Am. Ceram. Soc.* **33**(9), 274–283 (1950).
85. T. Chraska, A.H. King, and C.C. Berndt, On the size-dependent phase transformation in nanoparticulate zirconia, *Mater. Sci. Eng. A* **286**, 169–178 (2000).

86. R.C. Garvie, Stabilization of the tetragonal structure in zirconia microcrystals, *J. Phys. Chem.* **82**(2), 218–224 (1978).

87. A. Suresh, M.J. Mayo, and W.D. Porter, Thermodynamics of the tetragonal-to-monoclinic phase transformation in fine and nanocrystalline yttria-stabilized zirconia powders, *J. Mater. Res.* **18**(12), 2912–2921 (2003).

88. A. Suresh, M.J. Mayo, W.D. Porter, and C.J. Rawn, Crystallite and grain-size-dependent phase transformations in yttria-doped zirconia, *J. Am. Ceram. Soc.* **86**(2), 360–362 (2003).

Index

Printed in the United States
107667LV00003B/214-255/P